Competitiveness in the Food Industry

The Agricultural and Agro-Industry including Fisheries Programme (AIR) of Research and Technological Development was adopted on 23 April 1990 as part of the Community's Third Framework Programme. This specific programme concerns all of agriculture, horticulture, forestry, fishery, aquaculture and related food and non-food industries.

In relation to the food sector, the general aim is to provide, through pre-competitive RTD, the basis for environmentally-friendly and economically efficient processes (including transport and storage) for new or improved competitive products and a better knowledge of the characteristics, as requested by users and consumers, of final products derived from biological materials.

The programme is implemented through research and technological development projects, concerted actions, demonstration projects and accompanying measures such as research training grants and studies.

VISIT OUR FOOD SCIENCE SITE ON THE WEB

http://www.foodsci.com

e-mail orders: direct.orders@itps.co.uk

Competitiveness in the Food Industry

Edited by

W. BRUCE TRAILL
Centre for Food Economics Research (CeFER)
Department of Agricultural and Food Economics
The University of Reading
Reading, UK

and

EAMONN PITTS
Marketing Department
The National Food Centre
Dublin, Ireland

BLACKIE ACADEMIC & PROFESSIONAL
An Imprint of Chapman & Hall
London · Weinheim · New York · Tokyo · Melbourne · Madras

Published by Blackie Academic & Professional,
an imprint of Chapman & Hall, 2–6 Boundary Row, London SE1 8HN, UK

Thomson Science, 2–6 Boundary Row, London SE1 8HN, UK

Thomson Science, 115 Fifth Avenue, New York, NY 10003, USA

Thomson Science, Suite 750, 400 Market Street, Philadelphia, PA 19106, USA

Thomson Science, Pappelallee 3, 69469 Weinheim, Germany

First edition 1998

© 1998 Thomson Science

Thomson Science is a division of International Thomson Publishing

Typeset in 10/12 Times by Saxon Graphics Ltd, Derby

Printed in Great Britain by St Edmundsbury Press Ltd, Bury St Edmunds, Suffolk

ISBN 0 7514 0431 4

A catalogue record for this book is available from the British Library

Library of Congress Catalog Card Number: 97-74588

∞ Printed on acid-free text paper, manufactured in accordance with ANSI/NISO
Z39.48-1992 (Permanence of Paper).

Contents

8 New policies, new opportunities, new threats: the Finnish food industry in the EU 253
SAARA HYVÖNEN and JUKKA KOLA

9 Are Porter diamonds forever? 286
MAGNUS LAGNEVIK and JUKKA KOLA

Contributors

Steffano Boccaletti   Instituto di Economia Agro-Alimentare, Universita Cattolica del Sacro Cuore, Via Emilia Parmense 84, Piacenza, Italy

Chris van Egeraat   Department of Geography, St. Patrick's College, Maynooth, Co. Kildare, Ireland

Pat Enright   Department of Food Economics, University College Cork, Cork, Ireland

Xavier Gellynck   Department of Agro-Marketing, University of Ghent, Coupure Links 653, B9000 Ghent, Belgium

Saara Hyvonen   Department of Economics and Management, University of Helsinki, PO Box 27 (Vikki A-building), SF-000710, Helsinki, Finland

Jukka Kola   Department of Economics and Management, University of Helsinki, PO Box 27 (Vikki A-building), SF-000710, Helsinki, Finland

Magnus Lagnevik   Department of Business Administration, School of Economics and Management, Lund University, PO Box 7080, S-220 07 Lund, Sweden

Larry O'Connell   The National Food Centre, Department of Marketing, Dunsinea, Castleknock, Dublin 15, Ireland

Eamonn Pitts   Rural Economy Research Centre, Teagasc, 19 Sandymount Avenue, Dublin 4, Ireland

Helene Tjarnemo   Department of Business Administration, School of Economics and Management, Lund University, PO Box 7080, S-220 07 Lund, Sweden

Bruce Traill   The University of Reading, Centre for Food
                           Economics Research (CeFER), Agricultural and
                           Food Economics, PO Box 237, Reading RG6 6AR,
                           United Kingdom

Luciano Venturini   Instituto di Economia Agro-Alimentare, Universita
                           Cattolica del Sacro Cuore, Via Emilia Parmense 84,
                           Piacenza, Italy

Jacques Viaene   Department of Agro-Marketing, University of Ghent,
                           Coupure Links 653, B9000 Ghent, Belgium

Biographical details of contributors

Stefano Boccaletti, is an Assistant Professor in the Department of Agricultural and Food Economics of the Università Cattolica in Piacenza, Italy. His research interests include the economics of consumer behaviour, the analysis of structural change in the agro-food system and vertical coordination in the food chain.

Pat Enright lectures at the Department of Food Economics in University College Cork (UCC). A graduate of UCC and the University of Alberta, Canada, his research interests include regulation of the food industry, agricultural and food policy and rural development.

Xavier Gellynck is a researcher in the Division of Agro-Marketing, Department of Agricultural Economy, University of Ghent, Belgium. His main research interests are consumer attitudes and behaviour towards agricultural and food products, as well as competitive performance of the food industries.

Saara Hyvönen is Professor of Marketing at the University of Helsinki, Finland. Her research interests are strategy–performance relationships in food manufacturing companies, distribution channel management and marketing in SMEs.

Jukka Kola is Professor of Agricultural Policy at the Department of Economics and Management, University of Helsinki, Finland. His research interests include the development and welfare economic effects of agricultural and trade policy as well as the relationship between public policy and food economy.

Magnus Lagnevik is an Associate Professor in Business Administration at the School of Economics and Management, Lund University, Sweden. His main interest areas are internationalization processes and competitiveness as well as marketing strategies of food companies.

Larry O'Connell worked at the National Food Centre in Dublin, Ireland as a research officer during the period of the study. He is now a lecturer in the Department of Marketing at University College Dublin. His research interests include industrial competitiveness and the internationalization of business from peripheral regions.

Eamonn Pitts is Head of Food Marketing at the National Food Centre. In his research he has specialized for almost 20 years in issues facing the Irish and European dairy industries. He acted as co-ordinator of the sub-group on competitiveness within the EU Concerted Action on Structural Change in the European Food Industries.

Heléne Tjärnemo, MBA, is conducting research for her PhD thesis on ecological strategies in Swedish retailing at the Department of Business Administration, School of Economics and Management, Lund University, Sweden. Her main interests are marketing, retailing and the role of product quality in marketing.

Bruce Traill is Professor of Food Management and Marketing at the University of Reading, where he is also Head of the Centre for Food Economics Research (CeFER). He was leader of the EU research project Structural Change in the European Food Industries. His research interests include European integration and globalization as well as marketing strategies of food companies.

Chris van Egeraat studied economic geography at Utrecht University in the Netherlands. At the time of the research he was attached to the Economic and Social Research Institute in Dublin. At present he is pursuing a PhD at Maynooth College, studying the effect of international corporate reorganization on production linkages and industrial complex formation.

Luciano Venturini is an Associate Professor of Political Economy at the Faculty of Law of the Università Cattolica in Piacenza, Italy. His research activity includes the application of industrial economics to the agro-food industry with particular emphasis on the analysis of structural change and the role of non-price strategies.

Jacques Viaene is Professor of Agricultural Marketing at the University of Ghent, Belgium. His research interests are consumer behaviour and agribusiness competitiveness. Since 1996, he has been President of the Belgian Association of Agricultural Economists.

Preface

Competitiveness is widely discussed, in the media, by politicians, by businessmen and by economists. Regrettably they do not all speak the same language. There is considerable misunderstanding between these groups regarding the precise meaning of the term competitiveness. One major source of confusion arises because of the different levels at which competitiveness may be measured. Economists and politicians will frequently think in terms of the competitiveness of the whole economy. Businessmen focus on the competitiveness of their firm or sector.

This book is focused on sectoral competitiveness and uses the following definition of that term:

> a competitive industry is one that possesses the sustained ability to profitably gain and maintain market share in domestic and/or foreign markets.

This book of case studies in competitiveness has been compiled by a team of economists and marketing analysts based in six different EU member states. A grant from the Agro-industrial research programme (AIR) of the European Commission enabled the members of the group to meet regularly over a three year period in a Concerted Action on Structural Change in the European Food Industries. The concerted action was led by Professor Bruce Traill. The sub-group on competitiveness was led by Mr Eamonn Pitts.

Involvement in the concerted action enabled members of the group to debate the methodological issues and to present the results of their own work to critical review by interested colleagues. The individual chapters of this book, however, represent the work of the authors alone.

In this book, we review the various methodologies for measuring competitiveness and we show results of case studies of competitiveness of food industry sectors in six different EU member states. We ask how best to measure competitive performance and also how to explain that success or failure.

In the case studies the principal methodology used has been that of Professor Michael Porter. Porter analyses have been very widely used by business analysts in the 1990s. There is considerable academic controversy regarding the model derived by Porter, and the principal elements of this debate are discussed in Chapter 1. While we find some merit in the criticisms of Porter's explanatory model, we believe that the Porter approach to the analysis of the sources of competitiveness is the best available. This approach is essentially qualitative and multi-faceted, rather than econometric. In using it, we also take account of some qualitative factors derived from the literature on industrial districts and other sources.

The methodologies used are not, of course, specific to the food industry. We report on the empirical difficulties in practice of applying these methodologies in the food industry. These difficulties predominantly relate to the availability of comparable data and to the particular involvement of a supranational institution, the EU Commission, in policy affecting the food industry.

The Irish, Italian, and Belgian case studies are relatively straightforward applications of the Porter methodology in relatively successful industries. The Belgian case study uses in addition a development of the well known Boston Consulting Group methodology. The Swedish case uses the Porter approach to analyse an emerging market to see if an industry can reach international competitiveness in the future. The UK case applies a Porter analysis to an uncompetitive industry. The Finnish case analyses the role of different kinds of policy for industry strategies adapted to the situation in the agri- and food industry. In several respects therefore the collection of case studies here extends beyond the standard application of the Porter methodology.

We believe we have also made a contribution to the debate concerning the Porter model, particularly in relation to the role of 'Government', where there is a supranational institution, and in relation to the relative importance of co-operation as a component of competitiveness. The crucial role of retailers in the determination of the competitiveness of food industry sectors is also particularly evident.

Our overall conclusion is that Porter has provided a useful tool for the analysis of international competitiveness. In particular, it provides a checklist of items that should be considered and a guide to the ways in which they may successfully interact to create a competitive dynamic.

The method highlights the value of qualitative analysis applied through the case study approach. We hope that the results of our work will be of interest to colleagues in universities and research institutes and to consultants practising in the food industry in many countries. We believe also that our varied experiences will be a useful tool in teaching young researchers in universities and business schools. We are not aware of any other collection of case studies in competitiveness.

BRUCE TRAILL, Reading, May 1997
EAMONN PITTS, Dublin, May 1997

1 What determines food industry competitiveness?
EAMONN PITTS AND MAGNUS LAGNEVIK

1.1 Analysing competitiveness

The analysis of competitiveness is concerned with providing answers to classic questions of economics – what determines investment, what determines firm success and what government policies are optimal. 'Analysis of competitiveness seeks to address trade policy questions within the distorted world in which we live' (Abbott and Bredahl, 1994), replacing comparative advantage and relegating it to the status of a theoretical concept of little practical value.

Unfortunately there is no agreed definition of competitiveness. There has been an explosion of interest in the concept of competitiveness from a variety of viewpoints, in the past decade leading to considerable confusion in relation to the scope of the term. Competitiveness can be analysed at the level of the whole economy, an individual sector or an individual firm. As analysts of the food processing industry we are predominantly interested in measuring the competitiveness of the food industry and sectors of it. In our analysis we accept the 1991 definition of competitiveness from Agriculture Canada that 'a competitive industry is one that possesses the sustained ability to profitably gain and maintain market share in domestic and/or foreign markets'.

The definition relates to competitiveness in domestic and/or foreign markets. It is important for the analysis of competitiveness, that there are important relations between competitiveness in domestic and foreign markets. Even if the links are strong in open markets, there are no direct and self-evident connections between the two. As we shall see later in this chapter, the measurement of competitiveness in industries or sectors of the economy is often related to other sectors in the same domestic economy. In other cases, competitiveness in foreign markets is related to the performance of other imported goods to that market, not to the market as a whole. The role of foreign direct investment versus trade is another important question to which we will return. The basic idea is that there is no single measurement technique that covers all aspects of competitiveness. Combining various methods of relevance for the purpose of the measurement, will be the procedure that we recommend.

Competitiveness intuitively is a relative concept. One is concerned with performance *vis-à-vis* that of a competitor, whether firm or economy. Nowadays with dramatic changes in markets we need also to emphasize that it is also a dynamic concept, concerned with maintaining or winning market share into the future.

1.1.1 Measures of competitiveness

Buckley *et al.* (1988) have made a useful distinction between different measures of competitiveness:

- **Performance measures** look at how well a country, sector or firm has done relative to its rivals. Typical measures are profitability, growth, market share or balance of trade. **Revealed Comparative Advantage** (RCA) (see below) is a specific measure of performance developed and used primarily by economists.
- Measures of **competitive potential** look at the availability or quantity of inputs, which produce superior performance such as access to cheaper raw materials or superior technology, leading to price and cost competitiveness and higher productivity.
- Measures of **competitive process** are often qualitative in nature and seek to measure the management process or how competitive potential is converted into competitive performance. Examples of measures of competitive process are commitment to international business and marketing outlook.

In the first half of this chapter we are predominantly concerned with reviewing measures of competitive performance, and in the second half we focus on measurement of competitive process and potential, in particular the **Porter diamond** (Porter, 1990) and the industrial districts approach developed by Italian economists.

1.2   Measures of competitive performance

1.2.1   Revealed comparative advantage

If we wish to carry out an empirical study of international competitiveness in an individual industry such as dairy processing, and wish to forecast which of a number of trading partners are likely to specialize in it and what new patterns of trade will evolve after a process of economic integration, traditional economic theory (the **theory of comparative advantage**) has little to offer. We would need to have data not only on factor availability in this sector in each of the trading partners but also the factor endowments for all other industries as well, to ascertain where comparative advantage lies.

Refinements to the classical theory allow that a number of other factors, in particular scale of operations and technology, can have an influence in the process. But the economist's tools available to analyse comparative advantage are limited. There is considerable ability to analyse absolute differences in costs of production, but theory suggests this is of only limited value since it is 'comparative advantage' which counts. Buckwell *et al.* (1994) have pointed out that, despite ample data and much intellectual effort, there is little useful quantitative information on what proportion of EU grain production would survive in competition with North American and Australian producers in a free trade environment.

The ultimate arbiter of whether a country has a competitive advantage is whether, at the margin, it can produce and deliver products abroad at competitive prices without direct or covert government subsidies. 'This is usually something that can only be discovered through the operation of the market' (Buckwell et al., 1994).

To cope with this problem, Balassa (1965, 1977) developed a tool for measuring comparative advantage *ex post*, which he called **Revealed Comparative Advantage** (RCA). In our review of measures of competitive performance we are fundamentally concerned with the value of RCA and some of the variants which have been suggested since, as measures of competitiveness. The Boston Consulting Group approach is a similar methodology developed by business school analysts and is considered as an alternative.

Here, we use the term RCA only in the sense Balassa used it. However, in the recent literature the phrase is sometimes used to cover a wide range of trade-based measures (see for example Ballance et al., 1987; Vollrath, 1991).

Comparative advantage, Balassa claimed, would be revealed in relatively high share of export markets; comparative disadvantage in low shares of export markets. The RCA of a country for any particular good is the share of the international market for that good divided by its share of the international market for all goods. This fraction is often multiplied by 100 for ease of presentation.

An index of 110 for a particular industry in a particular country would mean that its share of the world market was 10% higher than its share in total exports and that the country had a (small) comparative advantage in that industry. Figures below 100 indicate comparative disadvantage.

The formula is expressed mathematically as

$$RCA = (X_i/X_{iw}/X_m/X_{mw})100$$

where X_i is the value of exports of commodity i from the country in question, X_{iw} is is the value of exports of commodity i from all countries,[1] X_m is the value of exports of all manufactured goods from the country in question and X_{mw} is the value of exports of all manufactured goods from all countries.

The RCA measure can therefore identify sectors for which an individual country has a comparative advantage and a comparative disadvantage. It measures relative success in exporting and (despite its name) is not dependent on any theory regarding interindustry trade, factor endowments, existence or other-

---

1. Data from a selected group of countries may also be used. Balassa (1965) confined his analysis to the 10 largest industrial economies. In some circumstances it would be perfectly legitimate to confine analysis to EU member states.

wise of free trade or perfect competition. RCAs are basic measures of success and failure and can provide useful data for which to test hypotheses in these other areas. Balassa himself (1977) used these indices to measure changing competitiveness of the US economy in research intensive industries.

RCA data can therefore be used as a measure of competitive performance as defined by Buckley. From our perspective, these data can identify successful and unsuccessful food industry sectors in individual countries. The only data required are trade statistics. The measure can be calculated for a whole industry such as food and drink, or for relatively small subsectors such as yoghurt, frozen fish, sausages or soup.

Trade data are available at a highly disaggregated level and it is usually possible to aggregate trade data to meaningful industry sectors.[2]

An RCA index, being based on trade data, can be calculated yearly, and trends in competitiveness in a sector identified by improvements or disimprovements in the level of the index. Although the focus is usually on performance within individual countries, by aggregating the data the technique can be used, to assess the competitiveness of sectors within large trade blocs such as the EU.

The absence of appropriate trade data, in general, precludes the use of RCAs at regional (i.e sub-national) level, though regional trade statistics are collected in some countries; for instance, the Bavarian regional government collects data on Bavarian trade including trade with other German Länder. The new method of collection of trade statistics within the EU (company rather than port level), introduced following the abolition of internal borders in, 1993, provides possibilities for wider collection of regional trade data.

1.3   National and international competitiveness

An RCA index tells us nothing about the competitiveness of the economy being measured. In a competitive economy total exports may be increasing faster than total world exports, and vice versa for an uncompetitive economy. These factors will not appear in an RCA measure, which is what it says, a measure of comparative advantage within each national economy.

An RCA index differs from a market share indicator in that the RCA index looks at relative or comparative performance of an industry compared to other industries in the same country, whereas the market share indicator looks at the absolute performance of an industry or company vis-à-vis its European or world competitors.

It is not appropriate to compare the level of an RCA index for the same industry across countries. If indices are measured simultaneously for a num-

2. This is not true for cheese where the trade classification on which the data are compiled lumps together many varieties. Only processed cheese and fresh cheese can be meaningfully analysed as sectors.

ber of countries it is legitimate to compare the trends over a period, as an indicator of what is happening in competitive terms or merely to identify for any individual sector those countries which appear to have comparative advantage in a sector and those which do not.

The reason why an RCA index cannot be compared across countries is that the absolute size of the index is affected by the size of the economy. Let us take as an extreme example a commodity where 100% of the world's exports comes from one country ($X_i/X_{iw} = 1$). If that country is small and accounts for 2% of the world's total exports (X_m/X_{mw}), the RCA for the commodity in question will be 5000 ($1/0.02 \times 100$). On the other hand, if the country is large and accounts for 20% of the world's total exports, the RCA for this commodity will be 500. In trans-European analyses of the food sector in Europe, Winkelmann *et al.* (1995) have found many very high RCA levels (> 1000) for Denmark and Ireland but none for the larger countries, for this reason.

Lange (1989) and Hartmann (1993) are therefore wrong to compile indices of competitiveness of the food industry in EC member states and of agricultural and food products worldwide, and of the food industry in EC member states, respectively, on the basis of the absolute value of RCA indices.

One minor problem with the use of RCA data for measuring competitive performance is that the data on the value of exports do not include subsidies received (e.g. export refunds in the case of food and agriculture exports of EU member states). The importance of these subsidies varies with the product type, being of far greater importance in the case of raw commodities than in the case of highly processed finished foods and more important for animal products than for fruit and vegetable products. Some EU member states (e.g. the Netherlands and Ireland) rely on third country markets for their exports to a much greater degree than others (e.g. France or Germany) and are therefore affected to a greater degree by this 'undervaluation' of their exports.

A similar problem may exist wherever export prices are affected by the internal pricing arrangements of multinational companies. These companies may decide (usually for tax reasons) to reduce or raise their prices artificially between subsidiaries. These prices will be reflected in the international trade statistics and will affect the levels of the RCA index. In the European food industry this is not likely to be a particularly widespread problem but there are some examples (e.g. cola concentrate from Ireland) where the export prices are believed to be artificially high.

Summing up, we find that RCA provides comparisons between industries within a nation, and that the measurement does not include subsidies and internal pricing adjustments within multinational companies.

1.3.1   Net export index

A major criticism of the RCA index is that it takes account only of exports and the level of imports is ignored. Balassa (1965) proposed an alternative measure which takes account of the net exports of a commodity from a country. Net exports are exports minus imports. If this figure is then divided by the total value of trade (exports plus imports) in the commodity in question, we have an index with upper and lower values of 1 and –1, which occur when there are no imports or no exports respectively. As with the RCA, these indices are usually multiplied by 100 for presentation so that the practical limits are –100 to 100.[3]

The net export index does not take account of the overall level of trade in a commodity. Only the relative value of exports and imports for the individual commodity are taken into account. There is no attempt to put these in the context of overall trade in the commodity. A country, which is relatively self sufficient, with a small exportable surplus and no imports, would have an NEI of 100 and thus appear to be extremely competitive even though it hardly trades at all.

Balassa recognized that (for his purposes) the NEI is theoretically preferable, but states that for an individual country imports are affected not only by taste factors but by various protectionist factors, which could artificially reduce the level of imports. If these factors were present, they might indicate a comparative advantage for an industry which artificially limited imports. This would obviously conflict with the theory of comparative advantage. However it would have no implications for the pure use of RCAs as measures of competitiveness, which recognizes the world as it is. Clearly an industry which is thus protected has an advantage.

In practice by Balassa found little difference between the two types of indices. Hartmann (1993), however, has discovered some quite wide variations and points particularly to the USA and Greece in her analysis of food industry competitiveness. USA appears much higher in the NEI than in the RCA measure, while the opposite applies to Greece. The probable reasons are that the NEI overestimates the performance of the USA, by not relating its exports of food products to its total trade, whereas the RCA measure for Greece is relatively high because small countries have high RCA levels in their advantaged industries.

A factor not considered by Balassa, which would affect the level of RCAs and indicate the superiority of the NEI, is the existence of entrepot trade. In Europe, the Netherlands is a major exporter of products which originate in other member states, are exported to the Netherlands and subsequently re-exported, usually outside the EU. The Netherlands also acts as the principal point of entry for commodities such as grain, spices and oils from outside the EU. It is probable, for these reasons, that the NEI should be used to measure the competitive performance of food industry sectors in the Netherlands. However, the new system of data collection on trade which has

---

3. Balassa subsequently suggests a refinement of this index, to apportion a trade surplus or deficit. This refinement affects only the level of the indices for different commodities and not their order.

been in use within the EU since January, 1993 will reduce this factor, known as the Rotterdam effect. Importing companies in other member states will now tend to report the true origin of imports, whereas the old customs-based system emphasized their most recent port.

1.3.2   Value added as a measure of competitiveness

Koutstall and Louter (1995) are critical of trade based measures of economic performance and propose an alternative measure. They criticize the trade based measures (including RCAs) used by Porter to define competitive industries on the following grounds:

- In a country with overall poor economic performance, some industries, although having a poor export performance in comparison to the same industries in other countries, will be denoted 'competitive'.
- If an industry supplies a large volume of intermediate goods to other exporting industries in its domestic market, this would not be taken into account in assessing competitiveness (ironically, in view of the emphasis in Porter's model on the importance of the home market).

In suggesting alternatives they state their preference for measures of economic performance which compare the same industries across countries rather than other industries within the one country. They also prefer total demand to foreign demand as this avoids what they call the 'unjustifiable influence of spatial scale' on performance of industries.

Their preferred measure of performance is **value added**. Static economic performance is calculated by dividing value added per inhabitant of a country by average value added per inhabitant in a group of countries. Dynamic economic performance is calculated as the percentage change in this indicator between two points of time.

These measures are then used to assess the competitiveness of the industrial sectors in 14 OECD countries. Indices are also compiled of the level of static and dynamic performance of broad sectors of Dutch industry (e.g. beverages and food). The published indices relate value added per inhabitant in these sectors to the average OECD level for all industries. Corresponding indices of 'food, beverages and tobacco' level are published for four countries (USA, Sweden, Germany and Netherlands). Clearly, if data were available, sectoral and sub-sectoral level indices of competitiveness could be compiled.

However, data on value added is generally not available in the same level of detail as trade data and the value added based index of performance suggested by Koutstall and Louter can only be used at a considerable level of aggregation, corresponding normally to three-digit NACE codes. For example, data on value added in the Irish dairy sector is only available for the dairy sector as a whole; separate data on cheese, fresh dairy products, butter or milk powders are not available.

A second problem is that data are compiled at enterprise level. The data of individual firms which have manufacturing activity in several sectors will be allocated to the sector with the greatest level of value added. The value added data for dairy processing may therefore include contributions from meat or animal feed, where these are manufactured by enterprises whose principal business activity is dairy processing. These problems do not arise with trade data.

The Koutstall and Louter value added index is a useful tool for measuring competitiveness, if the selected level of analysis is an aggregated industrial sector such as food. It has severe limitations, particularly in relation to data availability and quality, if applied at the more disaggregated level frequently required in analysis of sectoral competitiveness.

Within the EU, data may soon be widely available on the volume and value of output of specific products at a highly disaggregated level, as a result of the gradual implementation of the Prodcom Regulation (EU Regulation 3924/91). These data will give information on the volume and value of output (but not value added) of up to 5000 product categories, with the usual restrictions regarding confidentiality of information relating to individual companies. These data, when they become widely available, could replace trade data alone in analysis of competitiveness.

1.3.3   Boston Consulting Group growth share matrix

As we have seen, the RCA measure uses trade data and is therefore useful when analysing the competitiveness of an industry based in an individual member state. Much of the analysis of competitiveness that needs to be done is at firm level. The RCA measure is of little value here. However, in the literature from business schools a number of techniques for analysis of competitiveness have been developed, one of the most important and widely used of which is the growth share matrix devised by the Boston Consulting Group (BCG).

The BCG analysis that we introduce here is a development of the original company-based BCG analysis into an analysis of industrial sectors. This methodological development will be described after a short introduction of the original model.

The BCG's growth-share technique (Hedley, 1977) is based on a 2 × 2 matrix, high and low market growth rate and high and low market share. The basic technique was designed for evaluation by large firms of different businesses or products within its portfolio. Circles represent the current size and position of eight businesses making up a hypothetical holding company. The volume of sales of each business is proportional to the area of the circles. The location of each business indicates its market growth rate and relative market share.

The market growth rate on the vertical axis indicates the annual growth rate of the market in which the business operates. High growth markets are assumed to be more attractive because market share gains are more easily obtained. To distinguish between high and low market growth, 10% annum is considered as the midpoint.

On the horizontal axis, relative market share refers to the share of its market held by the business compared with that of its largest competitor. a relative market share of 0.1 means that the company's sales volume is only 10% of that of the market leader, while a figure of 10 would imply that its market share was 10 times that of its nearest competitor.

Relative market shares are drawn on a log scale so that equal distances on the graph represent equal percentage increases. The midpoint is 1.0; at this point, the company's market share is equal to that of its largest competitor. High market shares imply a strong competitive position.

Products or businesses can be classified into one of four quadrants.

- **Cash cows** are characterized by high market shares and low market growth. Profitability should be good; both investment and cash requirements are low. Such products should have a good competitive position and generate resources to support other product–market combinations.
- **Stars** are market leaders in high growth markets. Profitability should be good but investment requirements are high. These products should be top priority and become the company's future cash cows.
- **Wild cats** (also called problem children or question marks) are products with low market shares in high-growth markets. Cash requirements are high. If the position of these products cannot be improved, cash will be absorbed continuously.
- **Dogs** are characterized by low market shares in low-growth markets. Profitability is generally absent and cash requirements high. Divesting is normally recommended for these products.

The competitive position of two or more companies can be assessed by classifying their products in this matrix. Clearly firms with a large share of their sales in the cash cows or stars categories are more competitive than those with a large share in the dog or wildcat categories.

The BCG matrix is widely used for strategic market planning within large companies, with many business units, but has some weaknesses.

- Market growth is not a completely valid measure for evaluating market attractiveness. Other factors, such as entry barriers, bargaining power and size of the market are also important.
- Market share is limited as a sole measure of competitive strength. Other elements, such as location, degree of vertical integration and capacity utilization have an influence.
- Definition of the market is not always simple, as many market segments can be identified for one type of product.

The BCG matrix has the additional advantage that it can be easily used as an indicator of competitive strength, not only at the level of an individual company but also for an industry as whole. By mapping all product–market combinations of competing companies, competitive position on both domestic and export markets can be determined and compared with competitors.

Gellynck and Viaene (1993) have adapted this technique to enable them to analyse the competitiveness of the portfolio of products of an industry sector on foreign markets.

Product–market combinations are based on the combined nomenclature used in external trade statistics by Eurostat. Demand in the foreign market or markets is considered first and a matrix of the relative attractiveness of this market is determined for the portfolio of products of interest. On the supply side a matrix of the market position of the supplying country in these markets is determined. The competitiveness is the result of the interaction of the two matrices of attractiveness (demand) and position (supply).[4]

Only imports and exports are considered, for two practical reasons. First, comparable product statistics between countries are only available at import–export level and not at production level. Secondly, competition between exporters from one or different countries is considered much more relevant than competition between exporters and domestic producers. The model described below is developed for the demand approach. The **competitiveness** of an industry is measured by the positioning of its products in the refined BCG's 3 × 3 matrix, based on three levels of market attractiveness and three levels of market position.

The **market attractiveness** of a foreign market is determined by market size and market growth. For any product the size and growth in a market are compared with the size and growth of the market for other products. Market size is the volume or value of product imported into the examined foreign market in a particular year. Market growth corresponds with the percentage change in the volume imported into the examined foreign market over a defined period. For both variables a distinction is made between high, medium and low. Data can be collected for a number of national markets simultaneously.

The market size or growth is:

- **high** when the calculated value for the product is higher than or equal to the average of other products examined
- **medium** when the calculated value for the product is between the average and 50% of the average
- **low** when the calculated value for the product is lower than or equal to 50% of the average of other products examined.

The three levels of market growth and market size result in three levels of market attractiveness:

- **high** for products above the diagonal from top left to bottom right

---

4. In the original example the analysis was conducted on the competitiveness of Belgian processed meat products on the French market: this has subsequently been extended to other markets.

- **medium** for products on the diagonal
- **low** for products below the diagonal.

Market position on the foreign market is determined by market share and market share growth. The market share of a product from an individual country in an individual foreign market is the share of product from that country in total imported volume of that product in that foreign market in a particular year. Market share growth is the percentage change in the market share on the examined foreign market in the relevant period. For both variables a distinction is made between high, medium and low.

The market share or market share growth is:

- **high** when the calculated value for the product is higher than or equal to the average
- **medium** when the calculated value for the product is between the average and 50% of the average
- **low** when the calculated value for the product is lower than or equal to 50% of the average.

The three levels of market share growth and market share result in three levels of market position:

- **high** for products above the diagonal from top left to bottom right
- **medium** for products on the diagonal
- **low** for products below the diagonal.

For an individual industry with a varied portfolio of products, the results of market attractiveness and market position can be combined in a new 3 × 3 matrix which shows three levels of market attractiveness and position:

- **High** for the products above the diagonal
- **Medium** for products on the diagonal
- **Low** for the products below the diagonal.

In order to evaluate **competitiveness** of the exporting sector on the considered market(s), a similar market attractiveness/position matrix is created from the point of view of the exporting country. This means that the way supply from the exporting sector corresponds with and reacts to changes in demand from the foreign country is evaluated. When the exports have a similar or better market attractiveness/position than the imports the sector is considered to be competitive, and vice versa.

As in the case of the imports, market attractiveness and market position from the point of view of the exporting country can be combined in a new 3 × 3 matrix. It means that the competitive position of an exporting sector or country can be evaluated by:

- determining market size, market growth, market share and market share growth as defined above

- comparing the evolution on the importing market with the reaction of the exporting sector or country.

Table 1.1 shows the final output of the analysis of the competitive position of Belgian pigmeat industry products on the French market. (For each product–market combination, 'market attractiveness' is determined by market size and market growth. Market size in this example corresponds with the volume of product imported into France in 1987, and market growth with the percentage change in volume during the period, 1981–1987. Product–market combinations with both high market size and high market growth are considered attractive.

'Market position' is considered by evaluating market share and market share growth. The market share of each product in total imports of pigmeat is calculated for 1987. Market share growth is determined by evaluating the development of product market shares in total pig meat imports during the period 1981–1987. In the same way, the position of Belgian exports is determined in the matrix.

Table 1.1 illustrates the combination of market attractiveness and market position. It is clear that the Belgian pigmeat industry has a strong competitive position on the 'attractive' markets for carcasses and deboned meat in France. Belgium also occupies a strong position for shoulders, but this is a less attractive market. On the attractive market for ham, the position of the Belgian pig meat industry is weak.

Table 1.1 Market attractiveness and market position of the Belgian pig meat industry on the French import market, 1981–1987

Market attractiveness		Low	Market position Medium	High
High	F	Bellies **	Carcasses Deboned meat ***	Hams ***
	B	**	***	Deboned meat Carcasses ***
Medium	F	Shoulders *	**	***
	B	Hams *	**	Shoulders ***
Negative	F	Loins *	**	***
	B	Loins Bellies *	*	**

F, French imports; B, Belgian exports.
Market attractiveness/position: ***, high; **, medium; *, low.

The BCG technique is clearly designed to be used in strategic analysis at the individual firm level. It is dependent on data on the size of a market (world, regional or national) and on the market shares of at least two firms (the firm in question and its largest competitor).

The BCG techniques, being trade based, obviously have a lot in common with RCA indices, but appear to have the following advantages relative to RCA indices

- They are dynamic and take account of growth in the market. The RCA measure is essentially static. Although RCA levels can be calculated annually for an industry in a particular country, the trends discovered say nothing about the size or growth of the world market.
- They focus on competitiveness in a particular market. The RCA measure is a general measure.

Its disadvantages are as follows.

- The market is defined as the import market. It would obviously be useful, assuming such data were available, to include data on supplies from the home producer, which could in many cases be the dominant source of supply.
- The Gellynck–Viaene adaptation of the BCG technique in practice defines competitors purely in terms of other importers but it can be adapted for the whole market.

1.3.4 Foreign direct investment: the Dunning paradigm

The measures of competitiveness discussed to date have generally been trade based. However, nowadays 'foreign production' (production outside their home country by subsidiaries of multinational enterprises) surpasses food exports for many developed countries (see Balasubramyan, 1991 for the UK; Handy and Henderson, 1994 for the USA). Likewise, domestic production by the subsidiaries of foreign owned multinational enterprises is often greater than imports. As trade becomes a less important avenue for international transactions, questions are raised about the validity of measures of competitiveness based solely on trade data.

The authors mentioned above suggest that both trade and foreign direct investment are concentrated in the hands of the same very few firms. If a competitive industrial sector must be comprised of competitive firms, how should the activities of these multinational enterprises be incorporated into measures of national competitiveness?

Porter (1990) suggests that **home base** is the distinguishing characteristic:

As long as the local company remains the true home base by retaining effective strategic, creative and technical control, the nation still reaps most of the benefits to its economy even if the firm is owned by foreign investors or by a foreign firm!

Dunning (1977), on the other hand, defines a country's competitiveness as

the ability to supply its own and other country's markets through its own firms, wherever they are located.

In other words, ownership is the defining characteristic.

It is beyond the scope of this chapter to examine in depth the relative merits of these two alternative views. Traill and Gomes da Silva (1994) have developed a number of indices of comparative advantage, to assess changing competitiveness, which take account of foreign direct investment. These indices are basically adaptations of the RCA index already discussed, which give alternative values to the output of multinational enterprises both at home and abroad. They define several new indices which take alternative views on the role of foreign production in determining competitiveness.

The Porter-adapted index of revealed comparative advantage (PRCA) takes the value of exports and adds to it at national and world level the respective values of output produced by the country's outbound industry and by the world's outbound industry:

$$\text{PRCA}_i = 100[(X_i + \text{IPO}_i)/(X_{iw} + \text{IPO}_{iw})]/[(X + \text{IPO})/(X_w + \text{IPO}_w)]$$

These 'Porter adaptations' simply add foreign production to exports, giving it equal weight. Implicitly this assumes that all of a country's firms producing abroad retain their country of origin as home base.

The Dunning-adapted net competitive advantage index (DNCA) makes a different assumption. Like the Porter adaptations this measure gives output from foreign direct investment equal weight to exports. However the Dunning measure gives no value to domestic production by foreign firms, which is treated in the same way as imports. The DNCA is an adaptation of the NEI developed by Balassa and Bauwen (1988).

$$\text{DNCA}_i = 100 \ (X_i + \text{IPO}_i) - (M_i + \text{IPI}_i)]/(Y + \text{IPO}_i - \text{IPI}_i)$$

where IPI is the value of output produced within the country by inbound foreign direct investment.

Traill and Gomes da Silva (1994) carried out an empirical exercise to test the effects of making these adjustments compared with the conventional RCA measures. A major difficulty was the lack of data on foreign direct investment. The results indicated that the indices incorporating the output derived from foreign direct investment showed different levels from those based purely on trade, and in some cases different trends.

This exercise was to a degree theoretical to see what the effects of including 'foreign' production would be. There is clearly a problem in giving foreign production equal value to exports, but there is no agreed alternative measure.

The other major problem with this technique is its data requirements. The data on the value of output for the study were derived from asset sales ratios reported for the USA and from data on stocks of foreign investment, derived from various sources. The latter data are available only at a highly aggregated level (the food industry for example), and there are obvious problems with the general use in other nations (where the ratios of capital in final production could be quite different) of asset/sales ratios derived from the USA.

Where studies are carried out on a sectoral or product basis, data on industry output would need to be collected within countries on an individual firm basis. This may be possible for well documented sectors such as the dairy sector but would be much more difficult for other sectors.

In general, although the limitations of using trade-based data only in indices of advantage or competitiveness are recognized, there are considerable difficulties, both theoretical and practical, in seeking to incorporate the effects of foreign direct investment on the competitiveness of the home economy.

1.3.5 Measuring competitive performance in practice

What is the potential of these techniques for measuring the competitiveness of the European food processing industry?

As competitiveness is a comparative concept, we first need to know what it is being compared to. If the intention is to compare European to American or Japanese competition, we can say that the RCA measure, including adaptations of it to take account of the activities of multinational companies, could be a useful tool of measuring competitive performance, but there are problems in comparing the levels of indices.

The RCA will not measure competitive potential or competitive process as defined by Buckley. It is flexible in that it is possible to measure competitiveness at an appropriate disaggregated level. If we wish to use indices which take account of foreign direct investment, there are data and methodological problems to be solved but these are not insoluble.

The BCG technique is basically one for use within companies. However, the adaptation by Gellynck and Viaene has broadened its use to enable analysis of competitiveness of a sector within a particular foreign market. It could therefore be used to assess the competitiveness of European, American or Japanese industries in a third market, such as Mexico or Singapore.

To date, use of these techniques to measure the competitiveness of the European food industry as a whole has not been carried out. Studies have tended to concentrate on competitiveness of sectors or companies in individual member states. It is arguable in any case that for many sectors the member state or even the region is the appropriate unit of analysis, as there appear from the research to be regional or country-specific factors which contribute to competitiveness (see Fanfani and Lagnevik, 1995)

At the beginning of this chapter we defined competitiveness as being concerned with the classic questions of economics – what determines investment, what determines firm success and what government policies are optimal – and quoted a definition stating that analysis of competitiveness seeks to address trade policy questions within the distorted world in which we live. The measures discussed in detail in practice are only the starting point in any analysis of competitiveness. They help to define which are competitive sectors and which are not.

All of the techniques discussed here so far are measures of estimating competitiveness from past data on exports. They may not be good estimators of future or present competitiveness.

Fundamental questions regarding wealth creation and firm success and the lessons for public policy require the use of other techniques of analysis. These measures quantify but do not explain why. The measures of competitiveness discussed in the next section seek to answer that question.

1.4   Measuring competitive potential and process

In the analysis of competitiveness, the focus as we have noted can be on competitive performance, competitive potential or the competitive process. This section is primarily concerned with competitive potential and the competitive process.

There are at least five different research approaches in which the process of industrial development is studied in an effort to explain the sources of competitiveness.

- One is the **industrial networks approach** (Håkansson, 1992) which has its roots in international marketing, and where the main idea is that the company's competitiveness is developed in close interaction with other actors and organizations. The company competes with other organizations that could fulfil the same function in the network.
- Another approach with similar characteristics is the analysis of **agribusiness complexes** (ABCs), (Viaene, 1994) which consist not only of agricultural firms, but also of industrial and service firms and institutions supplying them with inputs and marketing their outputs.
- A third approach is **filière** analysis (Soufflet, 1990, 1994) in which the companies and institutions along the value added chain are analysed for different groups of products. Clusters and institutions are evaluated in the light of their contribution to the whole system.
- The two other approaches, which are subjected to more detailed analysis in this chapter, are the Porter (1990) approach and the industrial districts (ID) approach.

We examine the objectives and main results produced with these two approaches, as well as the methodologies used in the studies. We start with the Porter approach, since it is most widely used, and then examine the ID approach. We analyse the similarities and differences between the two approaches and we also discuss how the theory and methodology in the area of regional and strategic industrial competitiveness could be developed, using influences from the Porter diamond approach as well the ID approach. The comparative analysis is largely taken from Fanfani and Lagnevik (1995).

1.4.1   The Porter diamond approach

Porter is an economist who has for many years made important contributions to the thinking of business strategists. In this chapter we concentrate on his book *The Competitive Advantage of Nations* (1990), where he develops his theory of why particular industries become competitive in particular locations.

Porter's ambition was to write a 'book about why nations succeed in particular industries' (an ambition rather similar to that of Adam Smith). His aim was to help firms and governments to choose better strategies and make more informed allocations of national resources. But his idea is also to contribute to a more general understanding:

> What I am really exploring here is the way in which a firm's proximate 'environment' shapes its competitive success over time. Or, even more broadly, why some organizations prosper and others fail.

Porter also states that 'environment' means more than geographic location with its infrastructure and history. Other important factors are the way managers and workers are trained, the nature of the company's important customers, the nature of related and supporting industries and the role of national and local government.

The Porter approach looks at clusters of industries, where the competitiveness of one company is related to the performance of other companies and other actors tied together in the value-added chain, in customer–client relations, or in local or regional contexts.

Porter's analysis was made in two steps. In the first, clusters of successful industries have been mapped in 10 important trading nations. In the second, the history of competition in particular industries is examined to clarify the dynamic process by which competitive advantage was created.

With these steps Porter developed a 'general theory' regarding the factors leading to the creation of successful industrial clusters. These studies form a basis for the classification of nations into one of four stages of competitive development, reflecting both the characteristic sources of advantage of a nation's firms in international competition and the nature and extent of internationally successful industries and clusters. Conclusions on the upgrading of competitiveness are drawn and advice is given to nations and corporations.

Porter's book has had an enormous influence in two separate respects. His theory of the factors influencing upgrading of competitiveness is widely debated. Equally important, however, is the multifaceted approach to the analysis of competitiveness, including the **Porter diamond**, which has been adapted widely even by those who disagree with him on the relative importance of key factors. 'Porter analyses' have become a standard tool of the analysis of competitiveness, used by business consultants and academics alike. The other chapters of this book add to the literature and the appendix to this chapter lists other Porter analyses carried out on food industry sectors within the countries of the EU.

In practice, a Porter analysis of an industrial sector includes only the second of the steps in Porter's analysis. The mapping of clusters of successful industries was a necessary part of the development of his theory. It is not strictly necessary to carry out an analysis of any individual sector of industry, whether competitive or not. Most Porter analyses are simply concerned with the history of competition in a sector in order to explain its current competitiveness or lack of it.

The second step in Porter's analysis deals with the dynamic process by which competitive advantage is created. The basic method in these studies is historical analysis. The histories of the industries were based on historical sources and primary company documents. Interviews were conducted with current and former industry participants, industry observers and trade association personnel. Case studies were circulated in draft form to industry participants to solicit comments and additions. In some instances, case studies were also reviewed by experts from firms based in other nations.

The case studies were carried out so that they reflected both the country-specific aspects and the aspects of international competition in the same case. In some cases the structure is such that a comparison is made between industries in two countries. The phenomena that are analysed are classified into six broad factors incorporated into the Porter diamond, which has become a key tool for the analysis of competitiveness:

- **Factor conditions** are human resources, physical resources, knowledge resources, capital resources and infrastructure. Specialized resources are often specific for an industry and important for its competitiveness. Specific resources can be created to compensate for factor disadvantages.
- **Demand conditions** in the home market can help companies create competitive advantage, when sophisticated home market buyers pressurize firms to innovate faster and to create more advanced products that those of competitors.
- **Related and supporting industries** can produce inputs which are important for innovation and internationalization. These industries provide cost-effective inputs, but they also participate in the upgrading process, thus stimulating other companies in the chain to innovate.
- **Firm strategy, structure and rivalry** constitutes the fourth determinant of competitiveness. The way in which companies are created, set goals and are managed is important for success. But the presence of intense

rivalry in the home base is also important; it creates pressures to inno
vate in order to upgrade competitiveness.
- **Government** can influence each of the above four determinants of competitiveness. Clearly government can influence the supply conditions of
  key production factors, demand conditions in the home market, and
  competition between firms. Government interventions can occur at
  local, regional, national or supranational level.
- **Chance** events are occurrences that are outside the control of a firm.
  They are important because they create discontinuities in which some
  gain competitive positions and some lose.

The Porter thesis is that these factors interact with each other to create
conditions where innovation and improved competitiveness occurs. The
Porter diamond is illustrated in Figure 1.1.

The complexity of the approach becomes apparent when one reflects on
how to capture and measure all the phenomena mentioned in the model.
Since the competitiveness of a firm is dependent on its 'environment', in
order to get a full understanding we must

- analyse several companies in the same cluster
- understand the driving forces and key success factors in international
  competition
- have good insights into the relations between firms and governments.

1.4.2 Sources of competitive advantage

In the definition of factor conditions above we refer to specialized
resources specific for an industry and important for its competitiveness.
We need to develop further what these resources are and in what sense

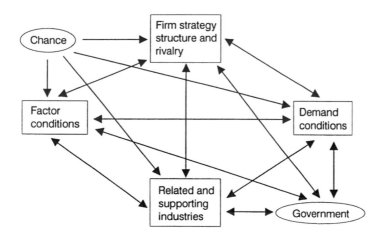

Figure 1.1 The Porter diamond (Porter, 1990, p. 127)

they are important for the industry and for the companies within the industrial cluster.

In doing this, we can portray the firm as a bundle of resources and capabilities. Resources include 'anything which can be thought of as a strength or weakness in a given firm' (Wernerfelt, 1984, p. 172). This resource-based view places more emphasis on the internal resources and competencies of the firm and less emphasis on the firm's performance in the market and working environment.

A firm can gain sustainable competitive advantage if its resources are

- **valuable** (it must be possible to use the resources to create value for the firm)
- **rare** (the supply of the resource must be limited)
- hard to **imitate**
- difficult to find **substitutes** for (Barney, 1991).

A resource-based view of regional clusters will point to the fact that there are resources that are external to any single firm, but internal to the region. (Herrigel in Kogut, 1993). Region-specific resources will lead to sustainable advantage when they fulfil the four criteria listed above. In many cases the resources of a region are based on unique historical conditions (Saxenian, 1985; Enright, 1994). In other cases the specific resources of a region are socially complex. It is fairly common in successful industrial districts to find complex relations between business and society, including family relations, ownership links, specific union relations, interaction between social and economic commitment and between government and industry. (Piore and Sabel, 1984; Pyke *et al.*, 1990; Fanfani and Lagnevik, 1995). For our purpose an analysis of key resources in successful industrial clusters will contribute to the understanding of upgrading of competitiveness.

In addition to understanding the nature of the determinants of competitiveness and the special characteristics of the available resources, we should also understand the dynamics – the change process whereby the competitive advantage is upgraded, sustained or lost – and we should understand the nature of the rivalry and the interaction between the different parts of the basic model. This means that we have a more complex model than those in which competitiveness is related to certain industry characteristics, such as size or market characteristics. Porter (1990) is convinced that this approach is useful:

> In studying national economic success, there has been a tendency to gravitate to clean, simple explanations and to believe in them as an act of faith in the face of numerous exceptions. The growing specialization of disciplines has only reinforced such a perspective. More can be done. Researchers in many fields of study are just beginning to recognize that traditional boundaries between fields are limiting. It should be possible to cut across disciplines and examine more variables in order to understand how complex and evolving systems work. To do so, mathe-

matical models limited to a few variables, and statistical tests constrained by available data, need to be supplemented by other types of work.

I have taken such an approach in this study. My theory seeks to be comprehensive and integrate many variables instead of concentrating on a few important ones. Sifting through over one hundred historical case studies is messy and not amenable to statistical analysis. Some will bemoan the judgements I made to chart national economies comprehensively. These choices reflect my conviction that understanding so complex and important a subject demands at least some research of this character....

In the dynamic process the key driving forces of competitive advantage are scale, cumulative learning, pattern of capacity utilization, timing of investment, level of vertical integration, location of the activity, institutional factors that govern activity, links between activities, the ability to share activities across business units and discretionary policies independent of other drivers (Enright, 1994, p. 9).

In the analysis of the dynamic it is also important to understand the technological development and the diffusion of technology. Another interesting phenomenon is the shift towards service activities of several kinds in the clusters. The level and quality of human resources is also a key strategic factor in the analysis of the triggers and driving forces in the agro-industrial clusters.

The firm that competes in the international market must configure and co-ordinate its activities. These activities that basically deal with localization and management of the firm relate to industrial clusters in that clusters must provide an environment to the firm in which it is easier to create superior buyer value or lower cost than in alternative locations. The advantages of the clusters derive from the fact that a number of activities can be more easily shared in the clusters: purchasing, warehousing, worker training, region specific promotions and lobbying, investments in environmental facilities and infrastructure, for example. Regional clusters often have a greater potential scope for sharing activities than geographically dispersed funds.

The approach is clear. Porter does not, however, prescribe a methodology. It is clear that a very large number of factors have to be taken into account by the analyst but there is no list of data requirements or statistical tests. A great deal is left to the individual analysts, including the depth to which they choose to go in the analysis. There are examples of Porter studies which take all the factors into account in 20 pages, while others take several hundred. There is no agreement on what precisely constitutes a Porter study.

Porter uses examples where he applies his diamond to a region, but there is no discussion in the book about the effects of regional applications of the model. Likewise the spatial definition of a home base is lacking. We learn in the book about 'tapping into other diamonds', but it is difficult to see the implications of this in Porter's own terms

Important contributions to the development of the Porter model have been made by De Man (1994) and Bengtsson (1994), especially regarding the concept of rivalry. Several authors have criticized Porter for using too narrow an approach when applying the diamond to the national level. Examples have been given from Canada (Rugman, 1993), Mexico (Hodgetts, 1993), and New Zealand (Cartwright, 1993). Dunning has also brought this up in a European perspective:

> Presumably, Porter would not wish to argue that the welfare of California was entirely, or even mainly, dependent on the competitive advantages of the location-bound resources of that state. Similarly – from 1st January, 1993, the analysis of Belgium or Ireland's national diamond, without considering their linkages with the assets and markets of the rest of the EC would surely be meaningless. Integration inevitably means national diamonds have to be replaced by supranational diamonds. (Dunning, 1993)

The essence of this discussion is the notion of the home base. The Porter diamond is criticized from a supranational perspective, where the argument is that the clusters must be viewed in an economically relevant context. Where national economies are linked to each other by agreements or unions, the argument is especially strong: the total context must be regarded as the home base, not only the nation in which the company is situated.

Ryan (1996) suggests that the appropriate solution in an EU context is to admit the EU and its institutions as a third exogenous force into the diamond, in addition to national government and chance. This appears particularly appropriate in analysis of food and agricultural sectors, in view of the central role of the Common Agricultural Policy (CAP).

Porter has left a number of questions to be answered by others, and in this chapter our ambition is to contribute to this discussion. Our method is to compare the approach with the ID approach and to discuss how useful the two are when applied to the agro-business and food industries.

1.4.3 Four stages of national competitive development.

Porter identifies four stages of national competitive development. The main sources for building of competitiveness are different in these stages (Porter, 1990, p. 543 ff).

- The **factor driven economy** represents a stage where the successful industries draw their advantage from production factors, mainly basic production factors. The economy is sensitive to world economic development and changes in exchange rates, which affect demand and relative prices.
- In the **investment driven economy** firms invest to construct modern, efficient and often large-scale facilities equipped with the best technology available on the global market. In this stage, foreign technology is not just imported and used, but also improved upon. In the investment driven economy, advantages are drawn from improving factor condi-

tions, but also from firm strategy, structure and rivalry. Although the home demand in this stage usually is rather unsophisticated, a large home market still contributes to the success factors.

- In the **innovation driven economy** all the parts of the diamond are in dynamic interaction with each other. Consumer demand is growing increasingly sophisticated, and world class supporting industries are developed. In this stage it is not relevant to talk about adapting foreign technology and improve it, but rather about creation of technologies. The range of industries involved is usually increased and the importance of competitive domestic service industries is increasing.

- The **wealth driven economy** is driven by the wealth that has already been achieved. At this stage the motivation to develop and innovate can deteriorate and the economy loses the ability to maintain its wealth. In wealth driven economies we find a large number of mergers and acquisitions. When companies are bought by established firms in an early stage of the life cycle, this can reduce the dynamism of the innovating sectors of the economy. The nature of investment changes from productive investments to financial investments.

1.4.4 The industrial districts approach

The ID approach is not as well known as the Porter approach. We therefore give a short introduction to it here, followed by a description of the theoretical and methodological points of departure. We also define the key elements in the analysis of competitiveness.

The ID method of analysis was originally developed in Italy. Many studies had shown deep differences in the level of development in different Italian regions. To explain the Italian experience, attention was concentrated on the local factors contributing to development in regions, where a strong concentration of small and medium-sized enterprises (SMEs) had grown up.

The regional and local studies in Italy utilized Marshall's concept which, almost a century ago defined an industrial district as the aggregation of many manufacturing SMEs in one place, with the advantages of the division of labour and economies of scale. Marshall's thoughts on the concentration of economic activities in space were a starting point in the studies of firms' strategies, market structures and industrial competitiveness (see Marshall, 1966).

The definition of industrial district was rediscovered and adapted by Becattini (1987, 1989) to the Italian postwar development after an in-depth analysis of several areas in which a strong concentration of industrial activities had taken place. He describes industrial districts as a localized thickening of interindustrial relations with reasonable stability over time. In order for a district to remain stable over time, there must exist 'a complex and inextricable network of external economies and diseconomies'. These factors are to be understood as external to the enterprise but internal to the region or the district.

The main characteristics of industrial districts are the close socio-economic relations between firms and families, which evolve jointly and dynamically, with a process of mutual adaptation to changing conditions. The geographic concentration and specialization of production in a limited area (county or few municipalities) has been accompanied by the concentration of independent SMEs, with some groups specialized in one phase of the production process and others specialized in specific final products.

Intercompany relations are in many cases based on co-operation and common interests, even if competition plays an important role. The most basic networks that characterize the industrial district are socio-economic and institutional relations. These networks are often based on common values shared by families and entrepreneurs, such as the value of work and saving, the risks propensity and the exchange of information and technologies. An important role is played also by the historical and institutional developments based on habits, co-operation and mutual assistance – collective services – but also general services in educational and professional institutions.

Sforzi (1987) found around 60 case studies of successful local development: other authors put the figure at around 100. The best known are textiles in Prato and Biella, knitwear in Carpi, jewellery in Arezzo and Valenza and ceramics in Sassuolo. The empirical studies show very large differences between firms, but also in productivity per worker within industrial districts.

Italian research work on local systems hardly touched the agro-food sector up to the mid 1980s, but, as a recent empirical study has shown, the agro-food districts could represent the main feature of recent change in Italian agriculture, in their process of integration and competition into the European market. The rapid and complex changes in Italian agriculture after the Second World War cannot be fully understood unless we pay attention to the process which led to the recent formation of agro-food districts. This process is closely connected to the change in Italian agriculture in relation to the entire food chain.

Research in the late 1980s stressed the geographic concentration of certain types of agricultural production and, above all, of their industrial processing (Iacoponi, 1990; Fanfani and Montresor, 1991, Fanfani, 1993).

These studies were based mainly on the analytical methods derived from the ID approach. This approach, which focuses on empirical studies with the aim of understanding how specific factors and actors co-operate and compete in a specific socio-economic context, has been used by researchers from many different disciplines, such as economics, sociology, applied statistics, business administration and political science. Since the theoretical background is pluralistic, it is not self-evident how the key concepts should be expressed, but the main thoughts behind the approach are described in Figure 1.2.

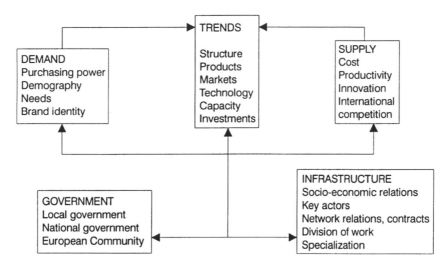

Figure 1.2 The industrial districts approach (after Viaene and Gellynck, 1995

The ID studies, in contrast to Porter analyses, start from the region, the companies and the society in the region. The method involves building up a good picture of the clusters. Thus it enlarges the concept of 'industry', since the studies not only describe the companies, but also analyse the infrastructure, consisting of services, supporting industries and socio-economic conditions. Interviews and questionnaires are used in all, or a sample of, the companies in a region. In these surveys, questions are asked, not only about a company's activities, but also how the company is related to other companies, local authorities, financial institutions and owners. The emphasis is clearly on understanding the company and the industry in a social and spatial context. In this respect the ID studies have a methodological advantage over Porter's studies.

As with Porter, the methodology is not clear although the approach is. There are no well defined steps in the analysis and much is left to the judgement of the analysts.

The analysis of the agro-food districts has shown close similarities with other types of industrial districts, not only in the specialization and concentration of production in a limited area, but also in their origin and in the internal organization of firms. At the same time, they clearly show some of the general limitations of district methodology.

The analysis of agro-food districts confirms one of the main strategic weaknesses entailed in the use of the industrial district as an analytical tool: i.e. that local systems cannot easily spread to other areas, since the conditions of their success can hardly be transferred (Amin and Robins 1990). In fact the ID concept implies that know-how, skills, traditions and co-operation, institutional involvement in specialized services and infrastructures exist that cannot be created by using the traditional instruments of regional policy. It is quite possible, however, to create models of the dynamic processes involved in the upgrading of competitiveness, and to create and to compare development patterns.

The ID approach has emphasized concepts that point out key areas of importance in the upgrading of competitiveness. These key concepts are:

- technological developments and their diffusion
- the progressive shift towards services,
- the level and quality of human resources
- the role played by institutions and local authorities.

1.4.5   Similarities and differences between Porter and ID approaches

Similarities. One similarity is that both approaches analyse industrial clusters. It is not regarded as sufficient to analyse an industry only on the aggregate level and use structural-behavioural explanations. In both approaches it is considered necessary to go beyond the traditional strategic analysis of the company. Competitiveness is determined not by company efforts alone, but also by environmental factors, other companies and public institutions.

As a consequence of the cluster approach, the definition of the task of the company is not taken for granted. Insourcing and outsourcing become important variables in the upgrading of competitiveness. Since participation in a network is regarded as an asset, it is important to use the options available in the network. If one's own company can gain competitiveness by letting another efficient producer take on parts of one's production, this is beneficial both for the other company and one's own. In the industrial districts, flexible specialization is an important asset since many companies can use a highly specialized subcontractor. On the other hand, the subcontractor needs orders from several companies in order to have sufficient volume. Without the network the strategy of the company would have to be redesigned.

Another common feature is that the relations between the company and the society are regarded as critical in order to increase competitiveness. In the Porter approach 'government' represents all kinds of factors that society can decide about and that influence the conditions for the business community. Porter claims that the most important production factors are the specialized production factors, often meaning national investments in earmarked R&D, provisions for venture capital and specialized education. Rivalry between competitors is also regulated by government. In the ID approach the relationship with society is very clearly expressed. Business is regarded as one part of the socio-economic life of the district, fully integrated with other parts. It is claimed that one cannot understand the business community without understanding its social setting.

A further important feature common to both approaches is that the industrial cluster is dynamic. It is not sufficient to measure and compare the competitiveness of a company, an industry or a region at a given moment in time. On the contrary, the main idea is to capture the dynamic forces that create change and improvement in the cluster.

Differences. The two approaches have different points of departure. In the ID studies the starting point is the industrial district, a limited geographic area where both business and social life are of interest to the researcher. How can the district or small region be developed and become prosperous? How can local resources be used in a constructive way? How can local companies increase their competitiveness? These are questions that interest the researchers in this tradition. Since the focus of attention is the region and since most of the companies in any region are SMEs, the special problems and possibilities of SMEs have caught the researchers' attention. Originally the focus was on SMEs only, but in recent research it is recognized that large companies and multinational companies can also be active in an industrial district.

Porter starts with quite another interest. He wants to understand the factors that drive the upgrading of competitiveness. He follows Schumpeter's tradition, as seeing the innovative process as central to the upgrading of competitiveness. Furthermore, he recognizes that there are important links between firms and their environment. In his search for driving forces he looks for extremely successful industries in different environments. Porter starts with industries that are competitive in the global market and he analyses the ability of large industrial nations to create environmental conditions that can stimulate innovation and drive the upgrading of competitiveness. The starting points of the two approaches are therefore contradictory.

They also have different theoretical points of departure. Porter is an economist: he states his problem in economic terms and even if he develops his study beyond the economist's traditional approach, he is very careful with his ties to economic theory. He relates to Schumpeter's view on the necessity of innovation, but puts less emphasis on the entrepreneur than Schumpeter does.

The ID tradition, on the other hand, starts from an intention to understand complex socio-economic interaction in a region. As a result, industrial districts have been studied by researchers from many disciplines, such as statistics, law, sociology, industrial organization, business administration and political economy. The multidisciplinary approach has also created several perspectives with complementary knowledge.

The multidisciplinary approach used in ID research suggests that it would be possible to study Porter's research problem using theory from other disciplines also. That would probably give new dimensions to the cluster analysis. In Porter's approach it is only the economic ties between the firms that define the cluster.

Global and regional competition. A cluster as defined by Porter is a national-level cluster engaged in global competition. The measure of competitiveness is based on the export share of the industry in question. This kind of data is available only on a national basis. The starting point of the ID analysis is the physical space, the geographic concentration of firms specialized in a specific industry. Italian experience shows that in many cases there could be two or more distinct industrial districts specializing in one national industry. These districts often differ in size, structure, and relationship with the foreign market. The Porter cluster lumps these different clusters together.

In Italy the production of ham and cured meats and provisions is concentrated in three main areas:

- Parma: 215 firms, 2500 workers, total production 1200 billion lire (18% exported)
- the San Daniele area (province of Udine): 26 firms, 700 workers, total production 700 billion lire (12% exported)
- Modena: 150 firms, 3000 workers, with production is concentrated on salami and other charcuterie products.

The Porter cluster for Italy puts all these different entities in the same cluster. In his analysis, 'pigmeat, dried, salted and smoked', is considered to be one national cluster with a world export share of between 9.1–18.2 %. In reality the production of dried meat, ham and salami is located in different regions with different structures, different technologies and different raw materials. Porter's analysis therefore does not provide a good understanding of the factors that promote and trigger competitiveness.

Another important industry concentrated in different districts in Italy is poultry, in which more than half of the Italian production is concentrated in two main areas: Forlì (Emilia-Romagna) and Verona (Veneto). These areas are completely different, but in this case also a Porter approach will unify them in the same cluster.

We argue, therefore, that a nation can consist of very different infrastructural characteristics and that the nation may be too wide an area to provide an understanding of the specifics of the upgrading of competitiveness.

Differences in methodology. In Porter's work the analytical framework is developed from the traditional economist's analysis of export and import data. His main point of departure is the definition of clusters with global competitiveness and their position in international industry. Important in the identification of significant clusters are national industry statistics, and the assessment of the role foreign investment plays in the national industry. In the analysis of the diamond, he has to use more qualitative methods and historical analysis to explain why dynamism occurs in the clusters. His method is basically interviews with experts, people who have significant knowledge of an industry and who can describe its development over time.

ID studies put more emphasis on the individual actors, and recognize the important role of entrepreneurs more than Porter does. Another difference is that in ID studies regional data are very important. Nations that have regional data for exports and imports are better equipped for studies of overall competitiveness in this research tradition. The availability of regional exchange data will facilitate a better understanding of competitiveness.

Differences regarding competition and co-operation. There is a considerable difference between the two approaches in relation to the role of collaboration and co-operation in improving competitiveness. ID studies have as a basic assumption that simultaneous competition and co-operation create good conditions for upgrading of competitiveness. Porter is crystal clear in his belief that competition is the driving force of economic development.

> This study, in a way I could not anticipate, has led me to a conviction that incentives, effort, perseverance, innovation and especially competition are the sources of economic progress and the basis for productive, satisfied citizens.

A number of other research traditions, however, indicate that co-operation is a very important ingredient in the upgrading of competitiveness.

> In a competitive environment where the actors are not atomistic, but exist within systems of actors, the relational capability could represent for entrepreneurial firms the way to gain sustainable competitive advantage. (Lipparini and Sobrero, 1994)

The importance of relationships between firms in the chain is also recognized in the literature on agro-industrial complexes and in filière analyses.

More recently, in the marketing literature on value chain analysis and efficient consumer response a central role is given to the management of chains and collaboration between companies in the chain as a means of ensuring competitiveness. Competition is seen as between chains rather than between companies.[5]

The same type of results have been found by the Uppsala School of International Marketing. Competitiveness is developed in close interaction with other actors and organizations in an industrial network. The ability to co-operate is important for the ability to compete. In this process, it is important to sort out core competencies and focus the company's activity on the functions where those are best used. If another company can perform a function better than our own, it is a competitive advantage for our company to co-operate with that company. This has also been found in Italian agro-food districts by Fanfani (1993):

> The great number of inter-firm relations show a segmentation of the production process, a development of services and 'contoterzismo' between firms, a deep specialization of firms in a certain type of product.

5. For a short introduction, see http:// www.fmi.org/media/bg/ecr1.html

According to Porter, domestic rivalry not only creates advantages, but also helps to avoid disadvantages. Fanfani and other ID researchers claim that it is also the domestic co-operation and interaction – horizontal integration – that creates advantages, or, most important, helps to avoid some disadvantages. This horizontal integration between firms and services creates advantages such as more flexibility in production and product innovation. Porter's basic assumption about competition as the only determinant of development creates an obstacle to a full understanding of the competitive processes.

In Svensson's (1993) study of the development of the Danish pork and dairy cluster it was shown that much competitive strength originated from strong co-operation. Porter's competitive postulates therefore form an unnecessary limitation.

Table 1.2 summarizes the main differences between the two approaches.

We conclude from our analyses that the Porter framework is a strong platform for analysis of the sources of competitiveness but that the ID approach offers additional concepts which can help us reach increased understanding of the competitive upgrading processes in industry. The approaches of the two methods are sufficiently similar that they may with profit be used in tandem.

In the case studies which follow the various authors have broadly followed a Porter approach in their analysis of the sources of competitiveness of individual sectors. However, they have absorbed some of the concepts of the ID methodology. Competition is not regarded as a key factor to the exclusion of co-operation.

The Porter diamond is an creative and powerful model which opens the door to a number of interesting studies of competition and competitiveness. In this chapter we have argued that Porter's empirical work does not provide material for the analysis that allows for full exploitation of the richness of the Porter diamond. The strength of Porter's work is that it gives an overview of the competitiveness of industries, starting from the global competition, and

Table 1.2 Differences between the industrial districts approach and the Porter (1990) approach

Aspect	Porter	Industrial districts
Point of departure	Global	Local
Theoretical base	Economic theory	Statistics, law, economics, business administration, political economy, sociology
Nature of cluster	National in a global context	Regional in an international context
Methodology	Economic context	Socio-economic context
Driving forces of competition	Rivalry	Rivalry and co-operation

it gives an interesting classification of a number of important industrial nations.

If we want to understand the functioning of an industry or analyse the reasons for competitiveness, we need empirical tools other than those presented by Porter. We have presented the methodological tools of the ID approach and argued that these could develop the analysis further. We cannot reach full understanding of the dynamics of competitiveness without breaking down the analysis to the regional level, getting better insights into socio-economic conditions and division of work in the value added chain, as well as a better understanding of the role of entrepreneurs, local and regional governments and unique competitiveness-creating actions.

Our main conclusion in this respect is that the Porter concept can be used on a regional level and that moving away from some of Porter's assumptions does not limit the potential of the concept.

1.5 Conclusion

Our initial aim was to review the techniques available for measuring competitiveness. We found the distinction made by Buckley *et al.* (1988) between competitive performance and competitive process and potential a useful one. Different techniques are required depending on the aspect of competitiveness of primary interest.

In the analysis of competitive performance, the RCA measure developed by Balassa is extremely useful at a national level. If measurement is required at firm level alternative techniques are required. The BCG growth share matrix is useful at firm level and has recently been adapted to allow it to be used at sectoral level. Both measures imply that success in exporting is a valid measure of competitive performance. As many firms now use foreign investment as an alternative to exporting in their efforts to win markets, this assumption may now be doubtful. Preliminary efforts have been made to incorporate the effects of foreign investment in measures of competitive performance. Value added indices meet some of the criticisms levelled at other measures but data is not normally available, except at a highly aggregated level.

When we came to discuss ways of measuring competitive process or potential, the dominant techniques available appeared to be subjective. Foremost among these is the Porter diamond, which is an approach rather than a methodology. This approach has already been subjected to widespread analysis and criticism, particularly in relation to its definition of a home base and its emphasis on rivalry rather than co-operation as a source of competitiveness. We found that the ID approach, developed in Italy, has many similar characteristics to the Porter approach but incorporates a number of additional elements. A combination of the two approaches would seem to offer the best solution to analysis of sources of competitiveness.

Appendix: Porter studies carried out in European food industry sectors

Estudio Sobre la Posición Competitiva del Sector de Alimentación y Bebidas en España. Series of 22 food sectoral studies and overall review published by Ministerio de Agricultura, Pesca y Alimentacion, Madrid, 1992. Studies carried out for the Ministry by Consultancy firm of Ernst and Young Asesores S.A.
International competitiveness in the Fruit Growing Industry. published by Rabobank Nederland, 1992. Study carried out by Agricultural Economics Institute (LEI) and Rabobank – reports on other sectors of the horticultural industry were also prepared by this team.
Kamann, D. and Strijker D. (1995) The Dutch dairy sector in a European perspective. in Beije, P.R. and Nuys, H.O. (eds) *The Dutch Diamond: The Usefulness of Porter in Analysing Small Countries*, Garant, Leuven-Apendoorn.
Lagnevik, M. (1996) *Svensk Livsmedelsnäring i internationell Konkurrens.* School of Economics and Management, Lund University, LOK Report no. 13.
Ryan, Ú.W. (1996) *The International Competitiveness of the Irish Shellfish Processing Industry – An Adapted Porterian Analysis.* Unpublished dissertation presented to University of Limerick in fulfilment of requirement of MBS degree
Svensson, M. (1993) *Struktur, Strategi och Konkurrens i de danska Mjölk- och Slakteribranschema.* Department of Business Administration, School of Economics and Management, Lund University.

References

Abbott, P.C. and Bredahl, M.E. (1994) Competitiveness: definitions, useful concepts and issues, in *Competitiveness in International Food Markets* (eds M.E. Bredahl, P.C. Abbott and M.R. Reed), Westview Press, Boulder, CO, pp. 11–35.
Agriculture Canada (1991) Task Force on Competitiveness in the Agri-Food Industry: *Growing Together: Report to Ministers of Agriculture.* Agriculture Canada, Ottawa.
Amin, A. and Robins, K. (1990) The re-emergence of regional economies? The mythical geography of flexible accumulation. *Environment and Planning: Society and Space,* 8(1), 7–34.
Balassa, B. (1965) Trade liberalization and 'revealed' comparative advantage. *Manchester School,* 33, 99–123.
Balassa, B. (1977) 'Revealed' comparative advantage revisited: an analysis of relative export shares of the industrial countries, 1953–1971. *Manchester School,* 45, 327–44.
Balassa, B. and Bauwens, L. (1988) *Changing Trade Patterns in Manufactured Goods: an Econometric Investigation,* North-Holland, Amsterdam.
Balasubramanyam, V. (1991), Entrepreneurship and the growth of the firm: the case of the British food and drinks firms. Lancaster University Discussion Paper EC8/91.
Ballance, R.H., Forstner, H. and Murray, T. (1987) Consistency tests of alternative measures of comparative advantage. *Review of Economics and Statistics,* LXIX, 157–60.
Barney, J. (1991) Firm resources and sustained competitive advantage. *Journal of Management,* 17, 99–120.
Becattini G. (ed.) (1987) *Mercato e Forze Locali: il Distretto Industriale,* Il Mulino, Bologna, p. 193.
Becattini G. (ed.) (1989) *Modelli Locali di Sviluppo,* Il Mulino, Bologna, p. 231.

Bengtsson, M. (1994) Climates of competition and dynamics – a study of interaction among competitors. Doctoral dissertation, Studier i företagsekonomi. Umeå University.

Buckley, P.J., Christopher, L., and Prescott, K. (1988) Measures of international competitiveness; a critical survey. *Journal of Marketing Management,* 4(2), 175–200.

Buckwell, A., Haynes, J., Davidova, S., Courboin, V. and Kwiecinski, A. (1994) Feasibility of an agricultural strategy to prepare the countries of central and eastern Europe for EU accession. Report to European Commission (DG I), December.

Cartwright, W.R. (1993) Multiple linked 'diamonds' and the international competitiveness of export-dependent industries: the New Zealand experience. *Management International Review,* 33(2), 55–70.

De Man, A.P. (1994) 1980, 1985, 1990: A Porter exegesis. *Scandinavian Journal of Management,* 10(4), 437–50. .

Dunning, J.H. (1977) Trade, location of economic activity and the MNE: a search for an eclectic approach, in *The International Allocation of Economic Activity: Proceedings of a Nobel Symposium* (eds B. Ohlin, P.O. Hesselborn and P.M. Wijkman), Holmes & Meier, New York.

Dunning, J.H. (1993) Internationalising Porter's diamond. *Management International Review,* 33(2), 7–15.

Enright, M. (1994) Organization and co-ordination of geographically concentrated industries, in *Co-ordination and Information. Historical Perspectives on the Organization of Enterprise* (eds D.M.G. Raff and N.R. Lamoreaux), Chicago University Press, Chicago.

Fanfani, R. (1993) Agricultural change and agro-food districts in Italy, in *Agricultural Restructuring and Rural Change in Europe* (eds D. Symens and A. Jansen), Agricultural University Wageningen.

Fanfani, R and Lagnevik, L. (1995) Industrial districts and Porter diamonds. Paper prepared within the EU Concerted Action Structural Change in the European Food Industry, presented at the Strategic Management Society 15th Annual Conference, Mexico City, 15–18 October 1995.

Fanfani, R. and Montresor, E. (1991) Il sistema agro-alimentare: filiere multinazionali e la dimensione spaziale dello sviluppo. *La Questione Agraria,* 41, 165–202.

Håkansson, H. (1992) Evolution processes in industrial networks, in *Industrial Networks A Net View of Reality* (eds B. Axelsson and G. Easton), Routledge, London.

Handy, C.R. and Henderson, D.R. (1994) Assessing the role of foreign direct investment in the food manufacturing industry, in *Competitiveness in International Food Markets* (eds M.E. Bredahl, P.C. Abbott and M.R. Reed), Westview Press, Boulder, CO.

Hartmann, M. (1993) Überlegungen zur Wettbewerbsfähigkeit des Deutschen Ernährungsgewerbes. *Agrarwirtschaft,* 42(6), 237–47.

Hedley, B. (1977) Boston Consulting Group approach to business portfolio. *Long Range Planning,* 10(1).

Hodgetts, R.M. (1993) Porter's diamond framework in a Mexican context. *Management International Review,* 33(2), 41–54.

Iacoponi L. (1990) Distretto industriale marshalliano e forme di organizzazione delle imprese in agricoltura. *Rivista di Economia Agraria,* 4, 711–44.

Kamann, D. and Strijker, D. (1995) The Dutch dairy sector in a European perspective, in *The Dutch Diamond: The Usefulness of Porter in Analysing Small Countries* (eds P.R. Beije and H.O. Nuys), Garant, Leuven-Apendoorn.

Kogut, B. (1993) *Country Competitiveness.* Oxford University Press, Oxford.

Koutstaal, M. and Louter. P. J. Measuring economic performance, in *The Dutch Diamond: The Usefulness of Porter in Analysing Small Countries* (eds P.R. Beije and H.O. Nuys), Garant, Leuven-Apendoorn.

Lange, D. (1989) Economic development and agricultural export pattern: an empirical cross country analysis. *European Review of Agricultural Economics,* 16, 187–202.

Lipparini, A. and Sobrero, M. (1994) The glue and the pieces, entrepreneurship and innovation in a small firm environment. *Journal of Business Venturing,* 9.

Marshall, A. (1966) *Principles of Economics,* Macmillan, London, p. 731.

Piore, M and Sabel, C. (1984) *The Second Industrial Divide.* Basic Books, New York.

Porter, M.E. (1990) *The Competitive Advantage of Nations.* Macmillan, London.

Pyke, F., Becattini, G. and Sengenberger, W. (1990) *Industrial Districts and Inter-firm Co-operation in Italy.* International Institute for Labour Studies, Geneva.

Rugman, A.M. (1993) The double diamond model of international competitiveness: the Canadian experience. *Management International Review*, **33**(2), 17–39.

Ryan, Ú. W. (1996) The international competitiveness of the Irish shellfish processing industry – an adapted Porterian analysis. Unpublished dissertation presented to the University of Limerick in fulfilment of requirement of MBS degree.

Saxenian, A. (1985) The genesis of silicon valley, in *Silicon Landscapes* (eds P. Hall and A. Markusen), Allen and Unwin, Boston.

Sforzi, F. (1987) L'identificazione spaziale, in *Mercato e Forze Locali: Distretto Industriale* (ed. G. Becattini), Il Mulino, Bologna.

Soufflet, J-F. (1990) Compétivité et stratégies agro-industrielles dans la filière viande bovine Européenne en construction. *Economie Rurale*, **197**.

Soufflet, J-F. (1994) Food quality policies and competition in the food chain: lessons for farmers. Paper presented at the 36th Seminar of the European Association of Agricultural Economists at the University of Reading, 19–21 September 1994.

Svensson, M. (1993): Structure, strategy and competition in the Danish dairy- and pork businesses. LOK report, Department of Business Administration, School of Economics and Management, Lund University.

Traill, B. and Gomes da Silva, J. (1994) *Trade, foreign direct investment and competitiveness in the European food industries*. Discussion Paper No 1, Structural Change in the European Food Industries, University of Reading.

Viaene, J. (1994) Agribusiness-complexes. Department of Agricultural Economics, University of Ghent.

Viaene, J. and Gellynck, X. (1995) Market integration and the small country case: pressure on the Belgian meat sub-sector. University of Ghent.

Vollrath, T. (1991) A theoretical evaluation of alternative trade intensity measures of revealed comparative advantage. *Weltwirtschaftliches Archiv*, **127**(2), 265–80.

Wernerfelt, B. (1989) From critical resources to corporate strategy. *Journal of General Management*, **14**, 4–12.

Winkelmann, M., Pitts, E. and Matthews, A. (1995*) Revealed Comparative Advantage in the European food industry*. Discussion Paper No 6, Structural Change in the European Food Industries, University of Reading.

2   Structural changes in the European food industry: consequences for competitiveness

BRUCE TRAILL

2.1   Trends in the European food industry

Despite the attention often paid to the glamorous high-tech sectors such as pharmaceuticals and information technology, it should not be forgotten that food and drink remains Europe's largest manufacturing sector in terms of value of output and is second in terms of employment, as shown in Table 2.1.

The food industry has seen dramatic changes in the last two decades. The completion of the Single European Market (SEM) has created a situation where, within Europe, food can be sold throughout the EU with little hindrance from tariff or non-tariff barriers. This means that competition is intensified, particularly for those firms which previously relied on national regulations to protect them from international competition, but opportunities are opened up for firms with international ambitions. These changes are not

Table 2.1 1994 ranking of EU manufacturing industries

Sector	Production (billion ECU)	% Production	Value Added (billion ECU)	% Value added	Employment (000 employees)	% Employment
Food, drink, tobacco	472.2	13.04	106.3	10.87	2330.7	11.54
Chemicals and man-made fibres	309.6	8.55	110.4	12.52	1584.4	7.84
Chemical industry	298.0	8.23	106.4	12.06	1539.6	7.62
Motor vehicles	276.3	7.63	80.7	9.15	1613.5	7.99
Electrical engineering	262.3	7.25	10.5	1.19	2334.0	11.56
Mechanical engineering	218.4	6.03	86.1	9.76	1981.3	9.81
Metal products	174.4	4.82	70.6	8.00	1986.5	9.84
Paper, printing, publishing	170.3	4.70	64.9	7.36	1350.0	6.68
Mineral oil refining	139.9	3.86	10.4	1.18	104.7	0.52
Preliminary processing of metals	119.4	3.30	32.3	3.66	662.3	3.28
Rubber and plastics	110.9	3.06	43.8	4.97	1060.3	5.25
Non-metallic mineral products	101.3	2.80	41.9	4.75	918.6	4.55
Timber, wood and furniture	77.9	2.15	27.8	3.15	846.1	4.19
Footwear and clothing	69.1	1.91	22.8	2.58	1020.0	5.05
Other transport	68.8	1.90	26.1	2.96	650.2	3.22
Office equipment	49.0	1.35	16.8	1.90	215.6	1.07
Other manufactured goods	n/a	n/a	24.2	2.75	n/a	n/a

Source: Estimates from *Data for European Business Analysis* 1995.

restricted to Europe: globalization is a trend affecting food, like other industries (Traill, 1997), and this trend is encouraged by the recently completed GATT round, by new technologies that make long-distance transportation of food products economically feasible, by a trend towards convergence in demand patterns among consumers world-wide, and by the emergence of new forces in global food markets, notably Japan. Dramatic changes are also occurring upstream and downstream of food manufacturing as agricultural policies are reformed in such a way as to become more market oriented, and as food retailing becomes more concentrated and more efficient, more international and consequently more powerful. Finally, consumer demand is evolving away from 'commodity' products towards more finely differentiated, high 'quality', value-added products. Manufacturers, to remain competitive in the modern world, must necessarily develop a capacity to innovate quickly and effectively. Most firms can no longer rely on producing a constant range of traditional foods. At the same time, policy-makers and industrial trade federations at various levels are concerned about the effects of these changes on 'their' firms and what they can do to provide conditions under which their firms can thrive. The various levels may be national (individual member states of the EU); supranational (the EU); or local (regions within a member state). Policies that might impinge on competitiveness include competition policy, R&D, food regulation, the setting up of conditions that favour the creation of networks and alliances, policies affecting inward and outward investment, agricultural policies, information strategies and policies that affect industrial structures.

The objective in this chapter is to set the scene for the case studies that follow by providing background data on the food industry in Europe and discussing some of the major trends in the industry, particularly as they relate to factors affecting food industry competitiveness.

2.2   Changes at the consumer level

Changes in food consumption patterns and food related behaviour are, to a large degree, responsible for the nature of the product changes that take place in the industry. Porter (1990) highlighted demand conditions as important innovation drivers which can lead to a nation's (or region's) firms gaining competitive advantage when demand is 'sophisticated' and a country (or region) leads a trend that others will follow.

2.2.1   Economic factors

Consumer incomes in the EU have been growing since the Second World War and have been converging as the poorer (mainly southern) countries catch up with the richer (mainly northern) ones. Incomes influence food consumption in several important ways. First, as Engel's law states, when incomes rise, the proportion of expenditure allocated to food declines. Although the relationship is not perfect, being influenced by national differences in preferences for food and by prices, it is amply demonstrated for the EU in Table 2.2. The richer northern European countries spend less than 15% of their total expenditure on food, whereas Greece and Portugal still spend more than 25%.

Second, although the proportion of expenditure on food falls, the level of expenditure on food nevertheless rises as incomes rise. Since consumers in the EU are adequately fed (they are not short of calories), they use the increase in income to upgrade quality (at a higher price) rather than increase the quantity of food consumed. The demand for quality may take a variety of forms:

- growth in consumption of convenience foods
- growth in meat consumption in the poorer countries
- growth in consumption of generally more expensive ecological foods
- more eating out and higher value-added foods in general.

This trend clearly offers opportunities for manufacturers to develop such value-added and often high-margin products.

Third, as incomes grow the diversity of food products consumed tends to increase and there are changes in food-related behaviour. For example, the relationship between the growth in personal incomes and the demand for

Table 2.2 Proportion of expenditure allocated to food in the EU, 1993

Country	GDP per capita (ECU)	Proportion of total consumer expenditure devoted to food (%)
Luxembourg	26 856	10.9*
Denmark	22 254	14.6
Germany	20 097	10.9
Sweden	18 256	14.4
France	18 640	14.5
Belgium	17 849	14.0
Netherlands	17 268	11.1
Italy	14 586	17.0
United Kingdom	13 835	10.9
Ireland	11 335	18.2
Spain	10 434	17.8*
Greece	7 406	28.3
Portugal	7 324	25.4

* 1991 figures.
Source: Eurostat. *and Agricultural Situation in the Community.*

convenience foods is related in part to changes in lifestyle such as going out more or working longer hours, as well as to increases in the economic means with which to satisfy this demand.

Fourth, although the importance of prices in determining consumption behaviour diminishes as incomes rise, relative prices of close substitutes are still important to consumers. Thus factor conditions within a country or region that affect the price of the final product are important for competition between that country or region and its competitors. Particularly important in this respect is the impact of reform of the Common Agricultural Policy (CAP) on agricultural raw material prices.

2.2.2 Consumer concerns

In recent years there has been increased concern amongst consumers over the wider non-economic aspects of food consumption; for example, diet and health, food safety, the environment and animal welfare.

In general, concerns over the relationship between food and health have increased across Europe, although there are clear differences between countries, particularly between north and south (Oltersdorf, 1992). This has been a major force for new product development in the food industry. Manufacturers have attempted to appeal to the health-conscious consumer with foods low in fat and sugar, high in fibre or made with beneficial bacterial cultures (e.g. probiotics in yoghurt).

A longitudinal survey of 'environmental consciousness' in Germany indicates that the number of consumers who are environmentally 'conscious' or 'active', increased from 37% in 1982 to 60% in 1991, though there are reports that environmental concerns in Germany are now diminishing. The number of consumers judged to be environmentally 'conscious' or 'active' in eight European countries ranged from 30% in France to over 60% in Germany, Sweden and Norway (Oltersdorf, 1992). Although concern often does not translate into modified buying behaviour, manufacturers are nevertheless responding to this trend through recycling and biodegradable packaging, organic and low intensity farming products, and so on. This aspect of change is highlighted particularly in the Swedish case study (Chapter 7).

Consumers have also become increasingly concerned over the way in which foods are produced. Many consumers have negative attitudes towards the use of certain new food technologies, and this has acted as a constraint to innovations such as the adoption of food irradiation, and has led to worries about reactions to modern biotechnology. Porter (1990) has pointed to the importance of a country setting regulations that lead international regulatory trends as another source of competitive advantage, but it is hard to assess, for example, whether the American liberal attitude to gene technology (at the time of writing, genetically modified soybeans and corn have been released) will give it a leading edge in associated industries or whether the cautious European approach will set a trend that enables European companies to build competitive advantage based on alternative technologies.

2.2.3 Demographic factors

The most significant trend has been the growth in participation of women in the paid labour force, which has increased demand for convenience products and contributed to the demise of the family meal in favour of 'snacking'. This has led to efforts to develop snack-style foods, particularly for teenagers, with an emphasis on health as well as convenience.

There are still significant differences in female participation rates across Europe: the rate in 1989 was 50.6% in Denmark, but only 25.1% in Ireland (Table 2.3).

Allied with, and in some cases facilitating, the increased participation of women in the work force, has been an expansion of household capital for the storage and preparation of food, including microwave ovens, freezers and food processors, though the incidence varies substantially across countries. In the UK, for example, 48% of households own microwave ovens compared with only 6% in Italy (Table 2.4). These appliances have facilitated quicker food preparation and led to the development of convenience foods with appropriate packaging (e.g. for microwaving straight from the freezer).

Ageing of the population, together with a growth in the divorce rate and postponement of marriage, has produced a reduction in average household size (Figure 2.1). Household consumption of many foods is directly linked to

Table 2.3 Proportion of women in employment in the EU, 1991

Country	Activity rate (%)
Belgium	30.0
Denmark	50.6
Finland	47.5
France	37.8
Germany	36.7
Greece	29.2
Ireland	25.1
Italy	30.0
Luxembourg	28.7
Netherlands	34.7
Portugal	38.3
Spain	25.6
United Kingdom	42.6

Source: Eurostat, *Europe in Figures*, Office for Official Publications of the European Communities, Luxembourg, 1993.

Table 2.4 Household capital ownership rates, 1990 (figures are percentages)

	Austria	Belgium	Denmark	Finland*	France	W.Germany	Ireland	Italy	Netherlands	Norway	Spain	Sweden	Switzerland	UK
Households with:														
Refrigerator	62	54	52	73	54	66	59	21	62	72	43	57	11	48
Combined fridge/freezer	42	53	50	31	50	43	43	83	60	34	51	50	86	53
Deep freezer	42	57	59	61	40	42	20	31	38	75	11	63	45	34
Any deep freezing	72	86	92	82	77	73	58	89	82	88	55	94	92	81
Food processor	45	38	16	51	56	49	25	30	29	41	3	55	31	29
Blender/liquidizer	31	37	36	44	16	24	35	14	14	42	23	12	14	38
Electric handmixer	71	48	78	75	52	68	22	24	57	48	54	45	45	29
Electric mixer (not hand)	33	41	38	–	64	27	41	49	35	46	15	26	40	30
Microwave oven	31	21	14	52	25	36	20	6	19	34	9	37	15	48

* 1994 figures.
Source: Food Marketing Handbook, UK.

the structure of the household, notably convenience foods once again, and package sizes.

2.2.4 The European diet?

Many of the factors influencing consumer food behaviour have been happening simultaneously throughout Europe and as a consequence there has, since the early 1960s, been a reduction in geographical differences in food consumption patterns.

Using the broad FAO product categories, Table 2.5 indicates the coefficient of variation in consumption across 29 European countries in 1961 and 1990. In all cases it is lower in 1990, indicating that the differences in consumption among countries have diminished.

It has been suggested that this is an ongoing process, culminating logically in the dreaded 'Euro-diet'. However, among many reasons why food consumption patterns should not be expected to converge completely among countries, even if socio-economic and demographic factors do, is that culture is an important influence on behaviour and cultural diversity has proved resistant to the pressures from foreign travel, global media and telecommunications. Another is that individuals differ, both within and between countries (see e.g. Steenkamp, 1996, for a general discussion of these issues). Different individuals have different 'values' which influence their food choices.

Recognizing that consumers within countries are not all the same, it is sensible to understand European food markets as 'groups of buyers that

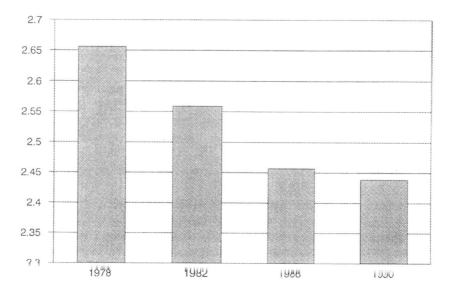

Figure 2.1 Decreasing European household size, 1978 1990 (EC and EFTA averages).

Table 2.5 Coefficients of variation of food consumption across 29 European countries

	1961	1990
Alcohol	70	52.5
Cereals	31.6	30.1
Eggs	47.3	31.3
Fruit	58.5	42.4
Meat	39.8	28.8
Milk	43.4	31.2
Pulses	99.1	80.5
Starchy roots	45	43.1
Sugar	41.4	21.8
Vegetables	43.5	42.6

Source: Computed from FAO food balance sheet data.

share the need and desire for a product and the ability to pay for it rather than those who share a common border' (Blackwell *et al.*, 1994, p. 221). According to this approach, demographic and economic considerations remain important, but so do psychosometric, attitudinal, cultural and lifestyle characteristics, and the process of convergence is best viewed as the growing importance of homogeneous segments of consumers which cross national boundaries. Such developments have substantial significance for trade in food products, which is dependent on the existence of consumers in one country wishing to buy the goods produced in another (see section 2.3.1 on internationalization).

2.3   Changes in food retailing in Europe

In northern Europe at least, long gone are the days when food supplies were bought daily from a variety of specialist local shops. Now, most foods are purchased during the weekly (or even less frequent) car trip to a supermarket or hypermarket and merely supplemented by local and speciality shop purchases. In western Europe as a whole, these two store types account for 96% of packaged food sales, 82% of soft drinks, 78% of cheese, 53% of 'fresh' fruit and vegetables, 42% of fresh meat, 34% of fresh fish and seafood and 30% of fresh bread (FMICC, 1992).

The incidence of super- and hypermarkets and of discount retailers varies widely among EU countries, as shown by Table 2.6. Once more a north–south divide is apparent, the north having more hypermarkets, supermarkets and discounters per head than the south which has more small independent retailers. This is also reflected in concentration ratios (Figure 2.2), which show the share of food sales going through the three largest chains, and are much higher in the north than the south of the continent. The indi-

Table 2.6 Food distribution in the EU, 1975–1991

Country	Super-markets[a]	Hypermarkets[b] (at 1 January)				Density[c] of hypermarkets	Market share of 'hard' discounters
	1990	1975	1981	1990	1991	1991	1990
West Germany	8000*	627	821	996	1004	1.3	22.0
UK	1950[d]	102[e]	279[c]	644[c]	733[c]	1.3	10.0
France	7050	291	433	790	849	1.5	2.0
Italy	3370	3	12	86	103	0.2	2.0
Spain	2500*	4	31	102	116	0.3	8.0
Netherlands	2050*	30	39	40*	x	0.3	6.7
Belgium	1919	70	79	98[f]	x	1.0	16.0
Denmark	944	5	x	49	x	0.9	12.0
Portugal	605[g]	4	4	18	20	0.2	x
Greece	5362[g]	x	x	18*	25*	0.2	x
Ireland	x	x	x	x	x	x	x
Luxembourg	51	3	3	5[f]	x	0.6	x

Countries are ordered by market size.
a, supermarkets 400–2499 m²; b, hypermarkets 2500 m² and above; c, number of hypermarkets per 100 000 inhabitants; d, between 460–2320 m²; e, more than 2320 m²; f, 1 January 1989; g, More than 200 m²; *, estimate; x, no information
Sources: Institute of Retail Studies and Institute of Grocery Distribution, *Economics and Statistics* No. 267, 1993–97.

cations are that southern European systems are becoming more like those in the north, though the speed of convergence is open to debate and depends, among other things, on national planning controls on out-of-town stores and competition policy rules on obtaining discounts from manufacturers.

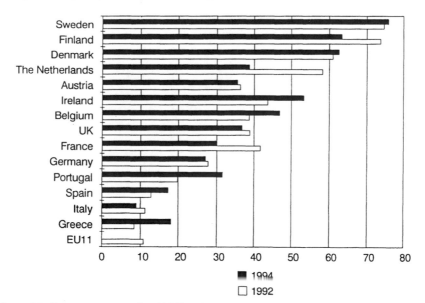

Figure 2.2 Grocery concentration (CR3) 1992 and 1994 (Rabobank, 1994; ABN Amro, 1996).

As retailing has become more concentrated, leading multiples have 'branded' themselves, trying to develop special images for quality, variety or value for money. They use technology to provide efficient distribution systems (often through their own centralized warehouses) which incorporate just-in-time delivery and enable retailers to hold minimal stocks (see Figure 2.3) but force manufacturers to innovate to find better and more flexible ways to cope with the less stable demand for their products that is the inevitable outcome. Information and market intelligence systems also use modern technology such as the well known barcoding and scanning systems, electronic ordering and invoicing between retailers and manufacturers, electronic price changing and transfer of funds, shelf space programmes and logistics programmes for the co-ordination of product flows. Combined with customer loyalty cards they give individual consumer purchase data that can be linked to postcode and other social and demographic data bases. This gives the retailers access to a wealth of consumer knowledge that can be used to design foods to target much finer segments of consumers. Such information is denied to manufacturers or is only available at very high cost and in inferior form (through market intelligence agencies such as Nielsen, AGB, etc.). The multiple retailers are thus generally thought of as the 'channel captains', who control the direction of the modern food system. This is reflected in their growing share of the profits of the food chain (Figure 2.4).

One of the main tools by which the modern retailer controls the food chain is **private label products**. Using their economic muscle combined with knowledge of the consumer, retailers are able to market their own private

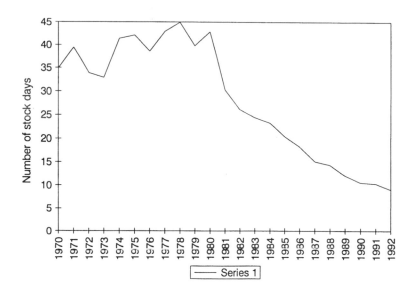

Figure 2.3 Tesco stores inventory 1970–1992.

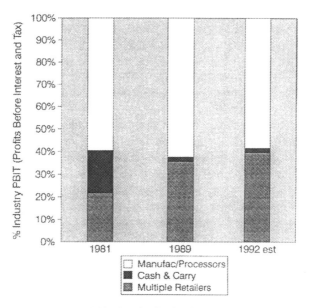

Figure 2.4 Shifting profitability within the UK food industry (ADAS).

brands of new food products that satisfy their consumers' needs. There are different types of private-label product:

- first and second generation products essentially provide cheap alternatives to established brands
- third generation products imitate the quality characteristics of established brands, maintaining a price advantage
- fourth generation products compete directly with established brands in terms of quality and innovation (Euromonitor, 1996).

In the UK, which is generally acknowledged to be at the forefront in fourth generation private label products, these retailer brands have often become the leading brands in the product category and generate considerable customer loyalty. In this sense, the retailer has become, in Porter terms, a sophisticated consumer, with very demanding quality specifications which could, in principle, impart competitive advantage to its manufacturing suppliers that they could exploit in other markets. This point is taken up in the UK and Swedish case studies (Chapters 4 and 7, respectively).

As private-label products rely on the store's own image, they require little advertising support and so are well placed to compete on price with the highly advertised branded products of the leading manufacturers. As far as the manufacturer is concerned, supplying private-label produce to retailers provides an opportunity for smaller[1] and medium-sized companies (SMEs) to operate in a market while avoiding the prohibitive costs of developing a

1. But not too small – they must be able to meet the retailers' volume requirements.

recognized brand. However, in the UK at least, most large companies also supply private-label products in addition to their company brands as this allows them to take advantage of economies of scale (as well as blocking their competitors' expansion plans). Only a few of the very best known manufacturer brands (e.g. Kelloggs) refuse to sell for private label, and use this fact as a promotional claim (implying that product quality is higher than can be obtained through private label).

Figure 2.5 shows the private label shares in a number of European countries. Even in Italy and Spain private label is becoming important, though whether levels anywhere will much exceed present levels in Germany and the UK is doubtful.

Although there has been a significant trend to Europeanization among the larger food manufacturers, through mergers and acquisitions, this trend has been far less apparent among retailers, most of whom operate solely in their own countries. Nevertheless, there is a trend towards internationalization, led by discounters such as Aldi and by French retailers which have been active in Spanish acquisitions (as well as some outside Europe). Another development has been the emergence of cross-country alliances of retailers in substantial buying groups (Table 2.7). These permit bulk buying and also the distribution of private-label produce throughout an alliance, though thus far the buying groups have not been very active. Much of the future development of the European food marketing system as a whole is surely dependent on the manner in which European food retailing develops.

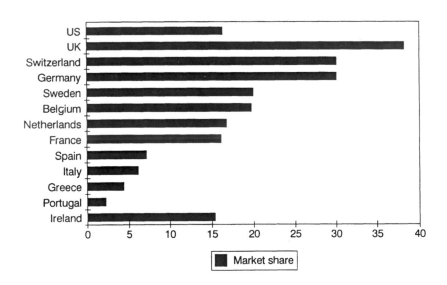

Figure 2.5 Market share of private label products 1992 (Rabobank, 1994).

Table 2.7 Main European retail alliances, 1992

Name	Members	Approximate size
AMS (Associated Marketing Services)	Ahold (NL); Allkauf (D); Argyll (GB); Casino (F); Dansk Supermarked (DK); Kesko (Fin); ICA (Swe); La Rinascente (I); Mercadona (Sp); Migros (CH)	>11% of the European market
CEM (Coopération européenne de Marketing)	Comad (I); Crai (I); Edeka (D); UDA (Sp)	ECU 28 bn
Deuro-Buying	Asda (GB); Carrefour (F); Makro (NL); Metro (D)	ECU 32 bn
Difra	Arlaud (F); Casino (F); Catteau (F); Coop Normandie-Picardie (F); Delhaize (B); Montlaur (F); Rallye (F); SCA Monoprix (F); Zanin (I)	ECU 18 bn
EMD (European Marketing Distribution)	Gelex (I); Markant (NL); Markant (D); Selex (Sp); Socadip (F); Uniame (P); ZEV (A)	15% of the European food market
ERA (European Retail Alliance)	Ahold (NL); Argyll (GB); Casino (F)	ECU 16 bn
Eurogroup	GIB (B); Rewe (D); Vendex (NL)	ECU 19 bn
Intercoop	Coop (I); Coop (CH); CWS (GB); EKA (Fin); FDB (DK); FNCC (F); KF (Swe); Konsum (A); NKL (N); Sok (Fin)	ECU 42 bn
Intercoop Trading	Despar (I); Spar (A); Spar (B); Spar (D); Spar (Sp); Spar (GB); Spar (NL)	Around ECU 17 bn world-wide
SODEI (Société développement international)	GIB (B); Pandoc (F)	ECU 8 bn

Source: Stratégies, 1992.

2.4 Changes in food manufacturing in Europe

The food manufacturing industry is composed of very diverse firms:

- giant multinational enterprises such as Unilever and Nestlé with representation in most food product sectors and turnovers measured in billions of ECU
- large companies with turnovers in tens or hundreds of millions of ECUs which tend to supply a limited range of national brands of processed foods and, increasingly, private label produce to national retailers, but do not, to any significant extent engage in exports or other multinational activity

- by far the most common category, SMEs supplying local or specialist (niche) markets.

Table 2.8 presents industry structural data for the EC (as it was) in 1990.

Looking at the proportions of enterprises in different employment size categories, it is apparent that more than 80% of food manufacturing companies employ less than 10 workers and 99.7% employ less than 500, which corresponded to the EU definition of an SME.[2] However, the remaining 0.3% of large firms employ 29% of the workforce and are responsible for 40% of turnover.

Writers on the food industry have often pointed to this very skewed firm-size distribution, arguing that the food sector is in some sense unique; Table 2.8, indicates that there are in fact no very substantial structural differences between food and 'all manufacturing' industries.

There are, however, persistent structural differences between countries and sectors, as demonstrated by Tables 2.9 and 2.10. Table 2.9 shows that the UK, Denmark and Sweden have high concentration industries (more than 50% of output from large enterprises); Germany and France have medium levels of concentration (30–50% of output from large enterprises); and Italy, Portugal, Belgium, Greece and Ireland have low levels of concentration (less than 30% of output from large enterprises). The Netherlands and Spain have insufficient data to categorize them in this way, but Table 2.9 suggests that the Netherlands has a medium level of concentration and Spain, a low level of concentration.

Table 2.10 shows the size distribution of the various sectors of the food industry, averaged over all countries. These data have been obtained from national data sources where the definitions often vary and they should not be regarded as having a high degree of accuracy. Three sectors have a relatively high concentration: dairy, brewing and chocolate. To these should be

Table 2.8 Size distribution of 'all manufacturing' enterprises and 'food, drink and tobacco' enterprises, 1990

Size class[a]	Enterprises (%)		Employment (%)		Turnover (%)	
	Total manuf.	Food industry	Total manuf.	Food industry	Total manuf.	Food industry
0	33.6	26.7	3.6	2.8	1.5	1.2
1–9	49.2	56.0	16.5	17.0	10.8	5.8
10–99	15.7	15.6	34.3	28.9	31.2	25.4
100–499	1.3	1.4	22.4	22.3	25.8	27.2
Total SME	99.8	99.7	76.8	71.0	69.3	59.6
500+	0.2	0.3	23.2	29.0	30.7	40.4

[a] Sizes are classified according to numbers of employees. '0' refers to a single self-employed person enterprise (zero employees); an SME is defined as a company with less than 500 employees.
Source: Eurostat, *Enterprises in Europe*, 1994

2. Changed in 1996 to less than 250 employees.

Table 2.9 Size distribution in the European food and drink industry in various countries – Eurostat data, latest year available

Country	Enterprises			Output		
Employees	0–99	100–499	500+	0–99	100–499	500+
Ireland*	86.5	12.8	0.7	34.8	56.2	9.0
Greece	89.6	9.4	1.0	31.2	41.3	27.4
Sweden	95.2	3.2	1.6	22.5	24.9	52.6
Denmark	95.4	3.5	1.1	21.3	18.9	59.8
Netherlands	96.7	3.3**	**	34.8	65.2**	**
UK	97.7	1.7	0.6	18.2	16.1	65.6
Portugal	97.9	1.9	0.2	35.5	38.1	26.4
Belgium	98.0	1.7	0.3	45.7	31.1	23.2
Italy	98.2	1.5	0.3	48.6	24.6	26.8
Germany	98.3	1.5	0.2	28.0	31.8	40.2
France	98.5	1.2	0.3	33.0	29.1	37.9
Spain	98.9	0.9	0.2	n/a	n/a	n/a
Weighted average	98.3	1.4	0.3	32.4	27.2	40.0

Source: Eurostat, *Enterprises in Europe* 3rd Report, 1994.
*, enterprises with more than three employees; **, medium and large enterprises combined.

Table 2.10 Size distribution of manufacturing sectors in the European food and drink industry

Employment	Enterprises			Output		
	0–99	100–499	500+	0–99	100–499	500+
Oils and fats	85.9	11.9	2.2	37.4	35.1	27.5
Meat	91.5	7.4	1.1	46.4	33.0	20.6
Dairy	86.0	11.9	2.1	13.2	31.3	55.6
Fruit and veg	85.2	13.4	1.3	37.3	54.6	7.9
Fish	91.4	7.4	1.2	52.0	36.3	11.7
Grain milling	95.5	4.4	0.2	71.7	24.6	3.7
Baking	95.9	3.7	0.5	55.8	29.0	15.2
Chocolate	86.3	10.5	3.1	24.1	35.8	40.2
Animal	94.7	4.8	0.5	56.9	34.6	8.5
Miscellaneous	89.1	9.1	1.8	28.6	42.4	29.1
Brewing	65.7	24.6	9.8	14.2	37.3	48.8
Soft drinks	87.3	11.0	1.6	31.3	37.7	29.3
Weighted average	92.2	6.7	0.9	38.6	36.4	25.0

Sugar is omitted because in many countries it is so concentrated that data are not released for confidentiality reasons. Starch and pasta have been included with miscellaneous foods as they are not reported separately in many countries.
Source: Various national statistical sources

added the most concentrated of all, sugar, for which data are unavailable for confidentiality reasons. These high-concentration sectors have the greatest economies of scale and scope for production, advertising and marketing, with greater involvement of multinational companies. Sectors with medium levels of concentration are the oils and fats, soft drinks, miscellaneous foods and meat sectors. They include sectors with dual characteristics, such as oils and fats and soft drinks, where giant multinationals (Unilever, Coca Cola) coexist with small local companies, and sectors which are in the process of restructuring such as the meat sector, where changing retail patterns are threatening the 'competitive fringe' (in this case local butchers). Sectors with low levels of concentration are the fruit and vegetable processing, fish processing, grain milling, industrial baking and animal feedstuff sectors. They cover both primary and secondary processing, having in common only the absence of economies of scale.

Somewhat surprisingly, an analysis of changes in industry structure during the 1980s, a period of some upheaval when the build-up to the completion of the single European market created waves of merger and acquisition activity, revealed very minor changes. There was no evidence to suggest a convergence of industry structures either across countries or sectors (Gilpin and Traill, 1995).

2.4.1 Internationalization of large food manufacturers

We have seen that large companies are responsible for more than 40% of the output of the food industry in the EC and more than 50% in some countries. Table 2.11 lists Europe's top 50 food manufacturers in 1994. The UK is the most common home base (12 of the top 20 and 19 of the top 50 companies originated there) followed by France, the Netherlands and Germany. The companies listed in Table 2.11 are mostly well known and one would anticipate them having a presence throughout Europe: surprisingly, a survey of the top 100 food processing companies by OC&C Strategy Consultants found half to have a presence in only one or two countries of the EU and only nine companies that sold in at least four of the five biggest markets (Germany, France, UK, Italy and Spain) (Financial Times, 1996). Thus there is a long way to go before the European market becomes truly integrated.

Market integration takes two main forms: markets may be connected either by international trade or by the establishment of multinational enterprises which undertake production activities in foreign countries. By both measures, European food markets are becoming more closely connected.

International trade has grown steadily since the end of the Second World War. Within the food and agricultural sector, the most rapid growth rates have been recorded for trade in processed products, a world-wide annual growth rate of 9.4% p.a. between 1961 and 1990 compared with 2.1% growth for bulk (agricultural) commodities over the same period. 'High-value' processed products now account for 85% of EU food and agricultural exports (Traill, 1997).

Within the EU, the most rapid growth has been in intra-Union trade. Most EU countries have also seen overall growth in the ratio of exports/imports, though the EU remains a net importer of food and agricultural products (Table 2.12). The nature of the trade has changed substantially. Before the

Table 2.11 Thr top 50 food manufacturers in Europe, 1994

	Company	Country base	Sales (US$M)	% Food and drink	Profit (US$M)	Main market	%	Main product	%	Empl (000s)
1	Unilever	UK/Neth	42 217	52	2944	Europe	54	Oils, dairy	30	294.0
2	Nestlé	Switz	40 770	96	2048	US	22	Dairy	27	209.8
3	Danone (BSN)	France	12 343	91	602	France	46	Dairy	36	56.4
4	Grand Metropolitan	UK	12 303	80	955	US	55	Drinks	45	87.2
5	Eridania/Beghin-Say	France	8 757	100	401	France	22	Oils	–	25.0
6	Allied-Lyons	UK	7 979	100	765	UK	62	Wine, spirit	38	25.8
7	Guinness	UK	7 065	100	1064	UK	20	Spirits	59	23.3
8	Hillsdown	UK	6 962	91	245	UK	42	Meal	16	43.3
9	Dalgety	UK	6 773	75	170	US	58	Distribution	56	15.4
10	Bass	UK	6 774	87	770	UK	87	Brewing	28	81.1
11	ABF	UK	6 645	72	512	EC	89	Milling	60	43.0
12	Saint Louis	France	6 210	31	126	France	36	Sugar	57	28.0
13	Tate & Lyle	UK	5 783	100	338	US	38	Sugar,Starch	89	15.8
14	Cadbury Schweppes	UK	5 644	100	630	UK	42	Beverages	55	39.1
15	Booker	UK	5 408	18	132	UK	92	Wholesaling	64	21.9
16	United Biscuits	UK	5 220	100	177	UK/Eire	44	Biscuits	48	36.4
17	Procordia	Sweden	5 088	25	616	Sweden	48	Beverages	11	40.1
18	Heineken	Nether	4 839	100	435	Nether	24	Beer	90	24.0
19	Sara Lee	Nether	4 014	65	320	Nether	35	Coffee	–	20.2
20	LVMH	France	3 813	50	529	France	47	Cognac,spirit	26	15.5
21	Whitbread	UK	3 555	100	268	UK	93	Drink,retail	100	35.0
22	Besnier/Bridel	France	3 373	90	71	France	76	Dairy	100	12.0
23	Südzucker	Germany	3 144	79	101	Germany	56	Sugar	73	11.0
24	Tomkins	UK	3 121	32	259	UK	50	Bakery	20	30.5
25	Northern Foods	UK	3 070	100	232	UK	–	Dairy	51	30.2
26	Campina Melkunie	Nether	2 996	82	19	Nether	53	Fresh,dairy	75	7.2
27	Unigate	UK	2 917	76	153	UK	82	Fresh foods	23	25.4
28	Oetker	Germany	2 901	43	–	Germany	50	Dry foods	–	12.3
29	Sodiaal	France	2 870	100	-6	France	86	Dairy	100	7.3
30	Bols Wessanen	Nether	2 756	85	180	US	49	Dairy	44	9.0
31	Pernod Ricard	France	2 552	100	199	France	54	Wines,spirit	55	9.4
32	Carlsberg	Denmark	2 388	100	212	Denmark	35	Beer	80	17.8
33	Orkla A/S	Norway	2 328	53	44	Norway	81	Beverages	16	14.7
34	Scottish & Newcastle	UK	2 294	59	277	UK	85	Beer	59	28.5
35	Fleischzentrale	Germany	2 216	100	–	Germany	90+	Meat	100	3.5

Table 2.11 *continued*

	Company	Country base	Sales (US$M)	% Food and drink	Profit (US$M)	Main market	%	Main product	%	Empl (000s)
36	Socpa	France	2 201	100	–	France	90+	Meat	100	4.5
37	Südfleisch	Germany	2 092	100	–	Germany	90+	Meat	100	4.5
38	Barilla	Italy	2 055	100	185	Italy	90	Pasta	55	8.2
39	Friesland Frico Domo	Nether	2 042	100	12	Nether	37	Cheese	–	5.6
40	Danisco	Denmark	1 995	55	126	Denmark	35	Sugar	–	11.5
41	Coberco	Nether	1 977	100	1 047	Nether	47	Cheese	32	4.1
42	Albert Fisher	UK	1 945	100	47	UK	36	Fresh prod	32	6.6
43	MD Foods	Denmark	1 905	100	95	Denmark	39	Cheese	43	6.1
44	Moksel	Germany	1 898	100	–75	Germany	60	Meat	100	2.6
45	Interbrew	Belgium	1 801	100	74	Belgium	62	Beer	–	9.8
46	Irish Dairy Board	Ireland	1 785	100	26	US	28	Dairy	100	2.4
47	Dairy Crest	UK	1 752	100	42	UK	91	Dairy	100	9.3
48	Bongrain	France	1 689	100	76	France	46	Cheese	35	7.5
49	Avonmore	Ireland	1 660	91	37	Ireland	27	Dairy	56	6.2
50	Altana AG	Germany	1 651	35	111	Germany	43	Dietetics	35	10.1

Source: UK Food Marketing Handbook.

Table 2.12 Trends in the ratio of exports to imports for agricultural and food products in value, 1985–1993

	1985	1990	1993
Belgium, Luxembourg	0.80	0.87	1.02
Denmark	2.26	2.20	2.13
France	1.22	1.34	1.38
Germany	0.47	0.51	0.58
Greece	0.86	0.72	0.81
Ireland	1.98	2.24	2.77
Italy	0.37	0.39	0.50
Netherlands	1.43	1.62	1.73
Portugal	0.59	0.40	0.32
Spain	0.92	0.82	0.87
UK	0.48	0.50	0.58
EU-12	0.57	0.63	0.77

Source: Eurostat, DG VI.

first World War developed countries imported raw materials, largely from their colonies, and exported manufactures. Now, most trade is intra-industry trade (IIT) between similar (developed) countries, the standard explanation being the exploitation of economies of scale in finely differentiated markets (Hirschberg *et al.*, 1994; Krugman, 1995). These are specifically the types of markets we identified earlier resulting from the emergence of cross-country segments of consumers with similar food consumption behaviour.

Within Europe, where levels of IIT are at their highest, they have also risen during the 1980s. Measured by the Grubel–Lloyd index, which ranges between 0 (no IIT) and 1 (where exports equal imports for each sub-sector

Table 2.13 Weighted average level of IIT in the EU food, drink and tobacco industries

	1980	1992
Belgium	0.57	0.62
Germany	0.53	0.58
Denmark	0.70	0.39
France	0.49	0.54
Greece	0.13	0.24
Italy	0.36	0.45
Ireland	0.32	0.38
Netherlands	0.54	0.56
Portugal	0.19	0.28
Spain	0.27	0.47
UK	0.45	0.40
EU12	0.38	0.45

Source: Gomes da Silva (1996).

of the industry in question), Table 2.13 presents estimates for 1980 and 1992 calculated at the four-digit ISTC level. They have increased in all but one country and have done so most where IIT was previously least important, in the Mediterranean countries (most notably Spain).

Although trade has grown strongly, the growth of international production (production by a multinational company's subsidiary in a foreign country) has been stronger. Data here are much harder to come by and less reliable, but it appears that foreign production by the subsidiaries of food multinational enterprises (MNEs) now exceeds the value of processed food exports by a factor of 4.3 in the USA (good data: Malanoski et. al., 1995) and is slightly greater for the total of a group of six major developed countries[3] (less good data: Traill and Gomes da Silva, 1996). In both cases, the growth in foreign production has been significantly greater than the growth in trade. Henderson and Handy (1993) suggest that the value of sales under licence now also exceeds the value of trade. The implications are becoming familiar: more competition for national firms and that competition is with companies that are generally considered to be world leaders.

2.5 Changes in policies affecting the EU food industries[4]

Recently, increasing attention has been paid to the linkages between certain public policies and competitiveness of the entire agri-food industry (Brink and Kola, 1995; Henson et al., 1995; Jones, 1995; Kola et al., 1996). The subject is also central to the Irish and Finnish Cases. The basic question for public policy in general is: how to guarantee that our country is an attractive location for firms of high international competitiveness?

Public policy can sometimes, however, result in negative effects, which arise mainly from attempts to isolate and protect some industry from competition. These attempts often lead to social costs, overall economic and international market distortions, and trade policy tensions. The intended beneficial redressing of the perceived market failure may turn into a policy failure. Positive effects generated by well-designed public policy measures should, in turn, be realized in enhanced competitiveness. Markets take insufficient account of non-market goods such as education, research and innovative activities required for the creation of the infrastructure (especially modern information superhighways) and provision of attractive factor conditions. Market forces are also unable to deal with environmental issues, which are becoming more and more important in all industrial sectors.

The main government policies affecting competitiveness of the food industry are industrial, competition, trade, investment, R&D policies and, last but not least, agricultural policy which, as operated in the EU under the

---

3. USA, UK, Germany, France, Italy and the Netherlands.
4. Jukka Kola kindly prepared the material for this section.

CAP has impinged on raw material prices for the food industry, the location of production (through quotas) and imports and exports of finished foods (through tariffs and export subsidies on their agricultural component), and has been responsible for much of the regulatory environment affecting food (e.g. geographical designations which deny Danish manufacturers the right to export Feta cheese within the EU).

Nowadays the approach of industrial policy is towards horizontal measures affecting all sectors equally, such as policies to promote SMEs, rather than measures which identify and promote specific 'key' sectors. Paradoxically, as Gassman (1996) points out, this development towards a more modest role for the state may involve a wider range of government institutions in order to effectively co-ordinate the diverse ingredients of industrial competitiveness policy.

Public policy can promote the gaining of competitive advantages, but it cannot create them. Through competition policy the public sector creates the environment, in which the firms build their competitive strategies. Porter (1990) argues that strong domestic competition creates competitive advantages, lack of competition weakens firms' international competitive position (see discussion in Chapter 1). Education and technology policies improve factor conditions, which are crucial in economic growth and international competition, including competition for foreign investment. Demand conditions of firms are strongly affected by, for example, taxation, trade, and more and more by environmental policy, which also influences production conditions. In addition, development of related and supporting industries and creation of firm networks can be encouraged.

The CAP strongly affects the functioning and operational environment of the European food industry and food chain. Various forms of price and income support measures have been applied to achieve the farm income goal. The results have often been poor: the farm income objective has not been reached, inappropriate support and regulation means have been used, structural development has been hindered, inequalities between regions and member states have been worsened, negative (environmental) externalities have been created, and excessive costs have been imposed on citizens through high consumer prices and taxpayer costs. Neither the 1992 reform of the CAP nor the 1995 Arctic–Alpine enlargement of the EU eased the situation.

The expert group of European agricultural economists published in late 1994 a study outlining the EU's agricultural policy for the twenty-first century (European Commission, 1994). At the very beginning they emphasized that unlike many policy measures and reform proposals in the past, their report has not been provoked by any acute crisis. Perhaps so, but one can argue that the CAP is in a continuous state of crisis. In addition, just as the 1992 reform was due as much to GATT Uruguay Round settlement on agriculture as to EU's own over-production and budget problems, the eastern enlargement had started to put pressure on further CAP reforms by the time the expert group started working.

The main elements in the expert group's reform proposal are:

- further cuts in support prices
- completely decoupled direct payments (to compensate for price cuts)
- a gradual phase out of the common (EU budget) financing of the compensatory payments
- elimination of quantitative production restrictions (e.g. quotas, set aside)
- better use of EC regional, social and cohesion funds for structural development.

2.6 Conclusions

There are many convergent trends in the food industry in consumption behaviour, retailing and manufacturing, yet there remain substantial differences between countries and between sectors and it will be some time, despite the closer economic integration of Europe, before a very homogeneous geographic pattern emerges. Nevertheless, some common driving forces exist throughout Europe, not least the rise of the multiple retailer, particularly in northern Europe, which has tended to neutralize (but not totally destroy) the power of the brand.

Large manufacturers have, naturally, made a number of strategic responses. First, they have seized upon the opportunities offered by the SEM to develop European as opposed to national brands and simultaneously increase their power relative to the retailers who remain predominantly nationally based (though, as we have seen, retailers too are slowly, 'Europeanizing' through mergers and the development of transnational buying groups). Second, manufacturers are searching for plants flexible enough to supply the wide product range demanded by fragmenting consumer markets. Third, they are reorganizing production into large plants capable of supplying the entire European market in certain product categories. Finally, they are seeking new products based on new technologies. That is, large firms are adopting a mixture of efficiency seeking (cost minimization) and product differentiation strategies to meet the changing circumstances.

Smaller manufacturers whose markets are national or local are also having to respond to change. They must either merge, develop products and strategies to supply Europe-wide niches or develop strategies which at least defend their home markets against foreign competitors.

Policy-makers concerned with encouraging the competitiveness of their domestic industries need to recognize first the way in which the market is developing and how this affects their market-relevant options. That has been the main object of this chapter. They must also recognize that not all firms are the same and they must be very clear about what aspects of the multifaceted concept of competitiveness they wish to encourage (for example, is the main goal to improve the trade balance, make the country more attractive to foreign companies, encourage the international expansion of their own multinationals, promote domestic income generating activities or develop the small firm sector?) and target their policy instruments accordingly.

References

Blackwell, R.D., Ajami, R. and Stephan, K. (1994) Winning the global advertising race: planning globally and acting locally, in *Globalisation of Consumer Markets: Structures and Strategies* (eds S.S. Hassan and E. Kaynak), International Business Press, New York, pp. 209–32.

Brink, L. and Kola, J. (1995) Small countries with large neighbours: choosing agri-food policies to improve competitive performance. Discussion group report, XXII International Conference of Agricultural Economists, Zimbabwe. *IAAE Members Bulletin* 13, May.

Commission of the EC (1994) EC Agricultural Policy for the 21st Century. DG for Economic and Financial Affairs. *European Economy, Reports and Studies* No. 4.

Euromonitor (1996) Private label in Europe: an overview.

Financial Times (1996) *Strategic Directions in European Food and Drink.* Financial Times, London.

FMICC (1992) Food retailing and distribution: issues and opportunities around the world. Food Marketing Institute (USA) and Coca Cola.

Gassman, H. (1996) Globalisation and industrial competitiveness. *OECD Observer* 197, 38–42.

Gilpin, J. and Traill, W.B. (1995) Small and medium food manufacturing enterprises in the EU: a cross-country synthesis. Discussion Paper No 11, Structural Change in the European Food Industries, University of Reading.

Gomes da Silva, J. (1996) Industrial structure and performance under economic integration: the case of the food industry. PhD. thesis, University of Reading.

Henderson, D.R. and Handy, C.R. (1993) Globalization of the food industry, in *Food and Agricultural Marketing Issues for the 21st Century,* (ed D I Padberg), FAMC, 93–1, Texas, pp. 21–42.

Henson, S., Loader, R. and Traill, W.B. (1995) Contemporary food policy issues and the food supply chain. *European Review of Agricultural Economics,* 22(3): 271–81.

Hirschberg, J.G., Sheldon I.M. and Dayton, J.R. (1994) An analysis of bilateral intra-industry trade in the food processing sector. *Applied Economics,* 26, 159–67.

Jones, W.D. (1995) Competition policy and the agro-food sector. Paper presented at the 44th EAAE Seminar, October 1995, Thessaloniki, Greece.

Kola, J., Hyvonen, S. and Vironen, T. (1996) Agriculture and food industry: convergence between public policies and business strategies, in *Agriculture and Rural Economy in a Large Integrated Economy* (eds K. Mattas, E. Papanagiotou and K. Galapanopoulos) Proceedings of the 44th EAAE seminar, Wissenschaftsverlag Vauk, Kiel, pp. 16–28.

Krugman, P. (1995) Growing world trade: causes and consequences. *Brookings Papers on Economic Activity,* 1, 327–77.

Malanowski, M., Handy, C. and Henderson, D. (1995) Time dependent relationships in U.S. processed food trade and foreign direct investment. Paper presented at NCR-182 Conference, Foreign Direct Investment and Processed Food Trade, Arlington, VA, 9–10 March.

Oltersdorf, U. (1992) Trends in Consumers' Attitudes Towards Food Quality and Their Influences on Food Consumption in Germany. FAO, Rome.

Porter, M. (1990) *The Competitive Advantage of Nations.* Macmillan, London.

Rabobank Nederland (1994) *The Retail Food Market.* Rabobank, Netherlands.

Steenkamp, J.-B. (1996) Dynamics in consumer behaviour with respect to agricultural and food marketing, in *Agricultural Marketing and Consumer Behaviour in a Changing World* (eds B. Wieringa, G. Grunert, J.-B. Steenkamp, M. Wedel and A. Van Tilburg) Proceedings of the 47th Seminar of the Association of Agricultural Economists, Wageningen, pp. 15–38.

Traill, W.B. (1997) Globalisation in the European Food Industries? *European Review of Agricultural Economics.*

Traill, W.B. and Gomes da Silva, J (1996) Measuring international competitiveness: the case of the food industry. *International Business Review,* 5(2).

3 A big industry in a small country: dairy processing in Ireland

LARRY O'CONNELL, CHRIS VAN EGERAAT, PAT
ENRIGHT AND EAMONN PITTS

3.1 The current state of the Irish dairy industry

The Irish dairy industry has grown rapidly since the 1980s against a backdrop of restricted supply of milk. The value of output increased over 100% between 1980 and 1990. Irish companies have been the third most active amongst European companies in acquiring dairy related international businesses, achieving this success through a strong focus on traditional commodity products especially butter and skim milk powder. The five largest dairy companies and the export marketing company (the Irish Dairy Board, IDB) remain small compared to some of their European counterparts. However a number of these companies are ranked in the top 20 dairy companies within Europe, in terms of size and profitability.

The importance of the dairy industry to Ireland made it a natural choice for the National Economic and Social Council to consider alongside two non-food industries in a study of the determinants of Irish industrial competitiveness. The aims of the study were to give advice to the the Irish government on how to upgrade dairy industry competitiveness and to find lessons of value to the Irish government for other industries. In the context of the European concerted action, the study provided an opportunity to apply and test the value of the Porter methodology and to assess the importance of clustering in Irish industrial development.

3.1.1 Scope of the research

The milk processing industry serves a basic nutritional requirement of consumers and provides ingredients for the wider food industry and other industries, such as pharmaceuticals. Figure 3.1 gives an indication of the range of products possible in the dairy industry and illustrates the complex inter-relationships that exist between the various product options. We examined those Irish companies involved in the primary processing of milk, without confining the analysis to any one product or group of products,

The processing industry is comprised of 92 independent entities two of which are foreign owned. However, the empirical research focused on 15 of the larger companies. The IDB is also included as it plays a major part in the export marketing of Irish dairy products.

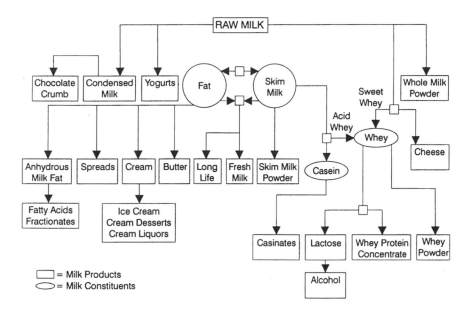

Figure 3.1 Milk products.

3.1.2 Research methodology

As a precursor to our study of the Irish dairy industry we first sought out and examined relevant articles, documents and reports. Before designing the field research we discussed the issues with a number of institutions and 'key informants' from universities, stockbrokers and bodies representative of the dairy industry.

We formed a view, based on interviews with key informants, that 15 companies were the most relevant to understanding the sources of competitiveness of the dairy processing industry. These were NCF, Mid West, Shannonside, Carbery, Nenagh, Tipperary, Lakelands, Golden Vale, Avonmore, Waterford, Kerry, Dairygold, Town of Monaghan, Wexford Creameries and the IDB. Of these, 11 were in a position to participate in the study.

In all 11 companies at least one in-depth interview with a senior executive was held, attended by at least two representatives of the research group. A structured questionnaire was employed with a number of closed but mainly open-ended questions. Only one interview was not recorded. We spoke to the Chief Executive Officer (CEO) in 10 companies.

A second questionnaire was administered at plant level. On most occasions this was left with a specific person to whom we explained the requirements for completion. The primary purpose was to provide us with more detailed information on customers and related and supporting industries.

Section 3.2 below provides a description of the evolution and performance of the industry. Sections 3.3–3.6 assess and explain the causes, of the degree of success which the industry has attained in terms of growth and international competitiveness by considering in turn the role of Irish factor conditions, domestic demand conditions, related and supporting industries and firm strategy, structure and rivalry as defined by Porter (1990). Section 3.7 looks at the role of the Common Agricultural Policy (CAP) as a potential exogenous determinant of the competitiveness of the Irish dairy industry. Section 3.8 draws together the conclusions and considers to what extent clustering in the sense used by Porter is significant for the competitiveness of the Irish dairy industry.

3.2   Evolution and structure of the dairy industry

3.2.1   History of the dairy industry

The development of the Irish dairy industry can be considered in four phases. First, a development phase, from the 1880s to the end of the First World War, in which the industry grew rapidly in a very competitive environment. Second, a phase of stagnation from the 1930s to the 1950s attributable to an international depression, domestic protectionist policies and a conservative dairy processing industry. Third, a period of renewed growth and reorganization as Ireland transformed its economic policy, prepared for and joined the EEC. The fourth phase began around the mid-1980s. It was characterized by the introduction of quotas which restricted the supply of raw materials while the larger Irish processors reorganized and have become international in scope, with four of the five changing to a co-op/PLC structure. Throughout the period since the 1950s, foreign investment by UK and US companies in chocolate crumb and powder manufacture played an important, though diminishing role, in upgrading technological and managerial skills in the industry.

3.2.2   Performance of the industry

The Irish industry performed strongly during the 1980s and up to the mid-1990s. Trade statistics demonstrate that Ireland has a revealed comparative advantage (RCA) in dairy products. In this section we seek to assess other measures of competitiveness compared with our main European competitors. A number of measures are employed to assess the competitiveness of the sector. However, there are a number of problems in using traditional measures of competitiveness in the dairy sector as growth in output is restricted by the quota regime and the market is heavily dominated and distorted by the presence of various support mechanisms. Further comparisons

Table 3.1 International comparative performance of the Irish dairy processing industry

	Ireland	Denmark	Netherlands	UK	France	Germany
Output (M ECU)						
1990	2710	2834	5004	7692	18 976	12 931
1980	1284	1531	5296	5709	9704	8372
% change[a]	+111	+85	−6	+35	+96	+54
Employment						
1990	9.5	8	18.3	38.8	60	39.7
Net change as % of 1980 Employment	−21	−16	−19	−20	−25	−20
Productivity						
Output per employee (00 ECU – 1990)	285	356	n/a	131	226	179
Growth in GVA m/employee over the period 77–90	x7.1	x1.8	x8.2	x3.1	x3.0	x2.4
Value added						
GVA m as % output – 1990	12.3	11.6		19.4	16.8	9.6
GVA M per employee	35	41.2		38.6	52.4	31.5
Profitability						
1990	6.2	2.8		9.9	2.6	3.8
1980	Negative	5.8		10.6	2.4	10.2
Investment						
% of output – 1990	3.8	2.3		2.9	2.8	3.2

Source: Based on Eurostat data and derived from Clarke, (1995).
[a] These figures are at variance with those reported by Kamann and Strijker (1995), but the trends and relative positions are broadly similar.

on profitability are distorted by variations in ownership structure. Notwithstanding these difficulties, a 'basket' of cross-country measures provides us with evidence of relative international competitive performance in the sector.

The Irish dairy sector has concentrated on increasing the level of output through a focus on cost-effective production of commodities. Table 3.1 compares the Irish dairy food industry to two other 'small' EU competitors (Denmark and the Netherlands) and three large competitors (the UK, France and Germany). In summary, Irish growth rates have been among the highest in Europe, as have productivity and the level of investment. Profitability has also improved strongly. At the same time employment has decreased, but broadly in line with the trend seen elsewhere. In terms of value added, Ireland remains somewhat below its competitors. In the following sections the individual measures are discussed in more detail.

Table 3.1 shows that the Irish industry achieved the greatest growth in value terms, between 1980 and 1990, with 111% growth over the 10 year period. However, this growth was achieved from a smaller base. In absolute terms, the increase in output is fourth highest, behind France and Germany but noticeably not far behind the UK – the difference is only 500 million ECU. The Irish companies have also been the third most active amongst European countries in international acquisitions (Perez *et al.*, 1994, p. 170).

When the productivity of the Irish industry is examined it is apparent that between 1977 and 1990 all of the countries examined have at least doubled the level of output per employee (Clarke, 1995, p. 29). The productivity of the Irish dairy sector has 'caught up' and is now second, behind Denmark and considerably ahead of the UK and Germany. In terms of value added[1] per employee Ireland has achieved second place in terms of growth behind the Netherlands over the same period. The level of value added, measured in terms of GVAm as a percentage of output, shows that Ireland is behind the UK and France but ahead of Denmark and Germany, with Denmark's position improving when the comparison is made on GVAm per employee. Clarke (1995) is critical of the sector because investment, although amongst the highest in Europe, has not been employed as effectively as elsewhere, namely UK, France and Denmark, to increase the proportion of value added.

The industry has improved in terms of profitability over the period 1980–1990. Clarke (1995) calculates sales margins[2] which indicates that prior to 1987 the Irish dairy sector recorded losses but subsequently margins steadily improved from 2.2% in 1987 to 6.2% in 1990, a level significantly ahead of competitors except for the UK.

Data on employment shows that job losses have been a feature of the food industry in Europe over the last decade. The percentage reductions by the Irish industry have been about average, with all competitors experiencing significant employment losses in dairying. Ireland and France recorded the highest cumulative percentage losses over the period 1980–1990 but only Denmark performed significantly better than Ireland. Further, Ireland recovered some of the jobs lost between 1990 and 1992. However, since then, the trend continues downwards with Clarke (1995, pp. 25–6) estimating 1994 employment to be 7400, down from 10 202 in 1989 and 12 000 in 1980.

In conclusion, although there may be some debate about individual measures, taken together they present the picture of a competitive industrial sector. The sector has concentrated on increasing the level of output through a focus on the cost-effective production of commodities with some sacrifice in the levels of employment. However, the rate of growth in the value of output has been among the highest in Europe, as has productivity and the level of investment, and profitability has also improved strongly.

---

1. Measured by gross value added at market prices. Eurostat gross value added at market prices (GVAm) is used, which includes net VAT, and this is defined as production value less intermediate consumption.
2. Measured as difference between output and intermediate consumption of goods and services (including labour) divided by output.

3.2.3 Industry structure

The industry is comprised of 92 independent entities, two of which are foreign-owned. It is dominated by co-operative and related forms of ownership. Co-operatives in Ireland process 99% of the milk supply compared to 48% in France and 92% in Denmark (Perez *et al.*, 1994). The industry can be divided into four categories on the basis of number employed (see Table 3.2).

- The first tier consists of five major companies, four of which are quoted on the Irish stock exchange and control over 75% of the milk pool. The combined turnover of these companies in 1994, including turnover of subsidiaries and non-dairy activities such as meat, was IR£3.9 billion, of which two-thirds was derived from international markets (annual reports, 1995). Dairy-related activities accounted for an estimated 67% (IR£2.5 billion).
- A second tier comprises nine companies, each employing 100–500 people.
- The third tier comprises 17 companies employing 21–100 people.
- The fourth tier consists of smaller companies, many of which are single owner/operator enterprises, e.g. farmhouse cheese manufacturers.

The industry is characterized by the presence of a large commercial international marketing board, the IDB, which is a commercial co-operative. Its function is to market products on behalf of its member companies (IDB, 1995). Figure 3.2 presents an overview of the structure of the industry and outlines the general relationship between the processors and the IDB. The extent to which companies use the IDB is dependent on their size and the price offered. In the main the IDB exports the products of the smaller co-ops and Dairygold and, to lesser extent, the produce of Golden Vale and Waterford. Kerry and Avonmore use the IDB more sporadically.

The IDB markets both Irish and foreign produce in international markets. It has an extensive world-wide distribution network, as well as packaging and manufacturing facilities in the UK and Belgium. The IDB does not have government authority to purchase milk supplies from the co-ops, unlike the New Zealand Dairy Board. Therefore the IDB can only influence the

Table 3.2 Categorization of dairy processors in Ireland by number of employees

Size (number of employees)	Total number of companies
< 20	60
21–100	17
101–500	9
500+	6 (includes IDB)
Total	92

Source: Company accounts (1994) and NFC Database (1995). (This is a private database compiled by the National Food Centre on Food processing companies in Ireland.)

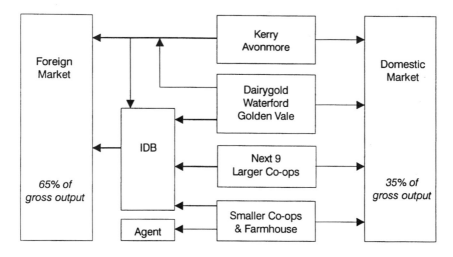

Figure 3.2 Structure of the Irish dairy processing industry.

strategic choices of co-ops through the commercial returns they offer, in competition with traders which cater for the smaller co-ops (interview with Mr Cawley, Chief Executive the IDB, 1996).

Foreign-owned companies accounted for less than 2% of the milk pool in 1994, down from 15% in 1980 (Forbairt, 1995). The foreign companies are involved in the manufacture of branded products, mainly baby food, and had exports of IR£190 million in 1994. However, although Forbairt (1995) have emphasized this as a key growth area, our research found that a number of companies did not consider this to be a practical strategy for Irish dairy processors. The world-wide baby food industry was considered by a number of respondents to be very concentrated and, further, the tendering system operated by developing countries was seen as an additional impediment to entry.

3.2.4 Location of dairy processors

The location of dairy processing companies is closely related to the production of milk. Proximity to a perishable and bulky raw material such as milk is a very important locational determinant. The early growth in the dairy processing industry resulted in the establishment of a creamery in almost every town and village in the dairying regions of the country. Improvements in processing and transport technology facilitated greater economies of scale. As a result of rationalization and amalgamation, especially in the 1970s, the industry gradually concentrated into a tiered structure.

Geographically, the result is an industry dominated by five large processors which are located adjacent to each other in the south of Ireland, in a band running through Munster and south Leinster. Many of the larger

Table 3.3 Dairy products sector employment, 1994

Region	No. of firms	Employment	% of employment
East	8	785	17.4
Midlands	2	140	3.1
North-east	8	437	9.7
North-west	1	52	1.1
South-east	12	1473	32.5
South-west	12	1431	31.6
West	4	206	4.6
Total	47	4524	100.0

Source: Forfas, Irish Economy Expenditures Survey, unpublished data.

processors have more than one processing site in this region, while liquid milk plants are located in the main urban centres outside the region. The second tier of processing companies are divided between the north-east (the second dairying region in Ireland), the west (a region relatively new to dairy farming) and the south. Foreign-owned processors have also located close to their source of raw material. In one case, two foreign-owned processors, a chocolate crumb manufacturer and a milk powder manufacturer, are located on the same site as a Dairygold plant. In another, a foreign-owned infant food plant, a co-op/private jointly owned cheese plant and a co-op plant occupy the same site.

Employment data reflect this pattern. Table 3.3 shows that 64.1% of employment in the dairy products sector is located in the south-east and south-west regions. The east region accounts for 17.4% of employment while the north-east accounts for 9.7% (Forfas,[3] unpublished data).

3.3 Factor conditions

The principal factors which are used by the dairy processing industry and which might have a significant influence on its competitiveness are human, physical, knowledge and capital resources and infrastructure. Porter (1990, p. 77) distinguishes between advanced and basic factors and between generalized and specialized skills (see Chapter 1). The most significant and sustainable competitive advantage results from possession of factors for competing in an industry which are both advanced and specialized. This typology is applied to the factors identified for the Irish dairy processing industry. The factors are also classified as to their effect on competitiveness, on a continuum between positive and negative impacts. Figure 3.3 classifies the factors of production used in the Irish dairy processing industry and can be used as a guide to findings arising from this section.

3. Forfas is a government agency that has responsibility for industrial policy.

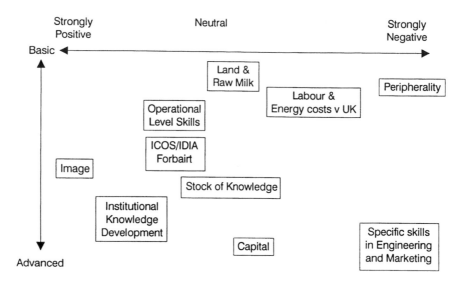

Figure 3.3 Impact of factor conditions and factor creating mechanisms on the dairy processing industry.

3.3.1 Physical resources

The most basic and general factors of production in dairying are the land and climate upon which dairy farming is dependent. Land of high quality, in tandem with suitable climatic conditions, allows the production of quality low cost milk which suggests that 'raw milk' is a general basic advantage for the processing industry, with a positive impact on competitiveness. In particular, it has allowed the industry to expand successfully as a commodity producer of milk products. However, there is an extra cost in the sense that the milk contains low levels of protein and fat and is produced in a pronounced seasonal manner. This suggests that 'raw milk' is better considered as a basic resource of the industry and the impact on competitiveness has been mild positive to neutral.

A second issue is whether Ireland's location has been a significant factor in the development of the Irish dairy industry. There is a cost disadvantage in serving export markets from Ireland. IBEC (1995, p. 22) estimate that it costs 150% more to deliver product to a UK customer from Ireland compared to a UK factory and that there is also a substantial cost penalty on deliveries to continental Europe. It might be expected that such a factor disadvantage would create extra pressure to develop efficient logistical systems. However a recent benchmarking study of eight European countries (DMBI, 1996) showed that while Ireland has a low cost level, measured by total warehousing, distribution and capital costs, an overall logistics index[4] places Ireland fifth in the sample of eight countries (Table 3.4).

4. Main elements of the index are lead time (50%), total costs in connection with warehousing, distribution and capital investments (40%) and safety stock levels (10%) (DMBI, 1996, p. 16).

Table 3.4 European logistic index for foodstuffs.

	Costs	Lead time	Safety stock	Overall logistics index
Ireland	54	145	133	108
UK	162	115	108	133
Sweden	79	97	200	100
Germany	82	109	57	93
Belgium	124	115	154	123
France	142	79	29	99
Denmark	83	79	67	79
Holland	78	61	51	67

Source: European Logistic Survey, March 1996, Danish Ministry of Business and Industry; survey performed by the Logistics Consulting Group.

The study concludes that this suggests that the 'money saved on direct costs is lost through reduced efficiency' (DMBI, 1996, p. 29). This poor performance is attributed to the long lead times (time from order receipt to delivery) in the industry and to high levels of stock, in part due to seasonality, carried in the industry (DMBI, 1996, pp. 30–1). Thus peripherality as a basic factor disadvantage has not materialized[5] to the extent one might have expected as a force for competitive innovation and improvement in the industry. Further, Cuddy and Keane (1990) suggest that peripherality may have caused firms to lag behind in technological dynamism and innovation, and concentrate on production at the mature end of the product cycle. We therefore conclude that peripherality has a mild to strong negative impact on competitiveness.

A third aspect of physical resources is the basic and general factor condition that is the image of Ireland as 'green and healthy'. This is a natural 'marketing' resource which is very important to all food producing companies. In a Porterian sense the image is a positive factor available only to Irish companies. It may be considered as a more important factor in consumer markets than in food ingredients markets where it is reported (PA Consulting Group, 1996) that nationality is the least important purchasing criterion. The impact on competitiveness has certainly not been negative, and in the successful development of the Kerrygold brand in Germany, it has been positive. Overall we conclude that it is mildly to strongly positive and that Ireland's image can be considered a created and reasonably advanced resource.

3.3.2 Human resources

An important factor in the dairy industry is the quality, availability and cost of human resources available to companies in the industry. It is also important to consider the 'factor creating mechanisms' available to the industry in developing the labour pool.

5. There are some examples, such as the use of ambient temperature techniques for the export of butter, but the overall index suggest that the impact of such innovations has been minimal.

This research found that the skill levels at operational level are considered excellent and that the companies have no difficulty in obtaining sufficient numbers of technical staff, with one company highlighting the availability of excellently trained operational level personnel as a key competitive advantage.

A number of specific deficiencies were mentioned, including engineers with specific dairy skills, experienced marketing people, especially international marketers with special knowledge of international channels of distribution, and commercially aware, rounded and broad-minded scientific people.

IBEC (1995, p. 18) estimate that labour costs (payroll and PSRI) represent 5.4% of processing costs. Thus labour costs, while important, are not a major aspect of overall processing costs relative to raw material. Further, although Ireland has a small labour cost disadvantage compared to the UK, it is not disadvantaged relative to competitors from other European countries.

Overall, the impact on competitiveness is not negative in the sense that Ireland is at least as well supplied as competitor countries with a good supply of advanced and basic labour skills which, while more expensive than the UK, do not constitute a major element of overall processing costs. It may be mildly positive at an operational level, with the supply and quality of more advanced graduate resources at least mildly positive.

3.3.3   Capital

In general Forbairt (1995, p. 65) feels the industry possesses good cash flow and has access to institutional funds, and has adopted innovative practices in its capital structure. It is difficult to discern the impact of capital availability, on competitiveness. However it is clear that the availability of capital was improved by the co-op/plc structure and the tax exemption on retained earnings. Therefore, this may be an enabling factor in the international expansion by Irish dairy companies. However, a number of interviewees suggested that a large co-operative, such as Dairygold, has equal access to capital. In conclusion we classified capital as an advanced factor which is not negative in its impact on competitiveness.

3.3.4   Knowledge

This section examines the proposition that Ireland is endowed with a knowledge base and knowledge creating mechanisms which facilitated the development of a successful Irish dairy industry. Technology is an advanced specialized factor which a recent report (Perez et al., 1995, p. 49) suggests is important in creating new product opportunities and allowing companies to gain competitive advantage.

Central to the competitiveness and upgrading of the sector is the role played by knowledge creating mechanisms in sustaining high international standards of technological knowledge in the dairy industry. A recent survey of European ingredient buyers found that Irish companies are perceived to be at least as technically competent as their competitors but they are not perceived to have a unique leading edge in any particular area (PA Consulting Group, 1996). Therefore, we consider that the level of technical knowledge available within companies had a positive impact on the competitiveness of the industry.

Forfas (unpublished data) report that expenditure on R&D outside their own company (non-internal) is small (IR£1.2 million) compared to the total R&D spend (IR£15.2 million) and further that this expenditure on 'non-internal' research is largely confined to the Irish market except in the area of technical consultancy. This suggests that the companies are sourcing most of their technology related services in Ireland. We found that companies considered research to provide a valuable service, especially in the area of basic product research and, to a lesser extent, in process research, which is at least as good as that of the main European competitors.

3.3.5 Role of state and other agencies

The state clearly plays a significant role in supporting the competitiveness of the dairy industry through its provision of a relatively strong education system (section 3.3.2) and research support (section 3.3.4).

The interviewees indicated that The National Food Centre (Teagasc) has played a positive role, especially in assisting in process and product development. However, some respondents felt that advice given to the farming sector was encouraging increased seasonality of milk production, thereby making it more difficult for the industry to adopt a more value added orientation. Although only recently established, Bord Bia, is important for companies which export independently as it provides market information and financial support. The Irish Dairying Industries Association (IDIA) and the Irish Co-operative Organization Society (ICOS) both fulfil the important role of providing information and lobbying at policy level in Ireland and in Brussels. The Department of Agriculture, Food and Forestry's positive function in co-ordinating, policing, administrating and lobbying is recognized throughout the industry. Forbairt was also considered to play an important role in the industry. In particular the Food Ingredients working group was cited, by a number of companies, as an example of good support.

Overall the support provided was in the form of technical advice and assistance as well as state grant aid in different forms. The impact on the industry is unlikely to be negative. The individual companies offered varied opinions, on this issue, but it is considered at least as good as that available in other countries in Europe.

3.3.6 Importance of factor conditions for competitiveness

Figure 3.3 summarizes the impact of the various factors of production used in the Irish dairy processing industry. Overall, the factor conditions are important for the competitiveness of the Irish dairy processing industry, particularly the level of technological knowledge, the operational skills and the supply of graduates, and the healthy national image. Factor creating mechanisms, such as the universities and the state support agencies and institutes, play an important role insofar as they help companies to maintain and update their expertise. These can be considered more 'advanced' factors, although some are quite generalized. Other strengths include basic factors such as land and raw material production. Peripherality is a basic factor disadvantage which has had a rather negative impact on the competitiveness of the sector. The lack of experience in international markets and specific skills in engineering were also cited as weaknesses.

3.4   Demand conditions

The market for dairy products can be broadly divided into a market for consumer products and a market for industrial products. Most of the consumer products are sold through retailers. Technological development has facilitated a greater use of milk constituents such as fat, protein and lactose, in a range of industries. Important applications of industrial dairy products are found in the animal feed, bakery, chocolate and confectionery, ice cream, baby food and diet food industries and in the dairy industry itself. Finally the fast-food chains and other forms of catering are becoming increasingly important customers of the dairy industry

Table 3.5 shows the relative importance of different markets for Irish dairy products. In 1990, 65% of output was exported.[6] Half of the sales to the Irish market are consumer products mainly sold through retailers. The local market for industrial dairy products is relatively small and the most important customer (36% of local sales) is the dairy industry itself, involving milk deals and the intraindustry sales of dairy by-products and ingredients for the production of yoghurt and processed cheese. Only 7% of the local output was sold to other industries, mainly beverages, confectionery, baby food, ice cream, bakery, meat processing, and ready meals industries. Many of these local industrial customers are subsidiaries of multinational companies.

Porter (1990) argues that proximity to the right type of buyers is decisive for national competitive advantage. Below we will examine the importance of local demand conditions for the competitiveness of the Irish dairy processing industry. Demand is examined, first, in terms of the individual dairy products and second, through an analysis of important customers in the Irish market.

3.4.1   Home demand for individual dairy products

The structure of home demand is seen as shaping the attention and priorities of a nation's firms which in turn influences competitive advantage in the production of certain products. Is it possible to identify segments that represent a large or highly visible share of home demand but account for a less significant share in other nations? Table 3.6 shows the *per capita* consumption for selected dairy products in 1993. The figures indicate that Irish people are by far the biggest milk drinkers in Europe. The *per capita* consumption of butter is below the EU average and consumption of cheese and yoghurt is amongst the lowest in the EU. Most other dairy products (skimmed milk powder, whole milk powder, whey powder, casein/caseinate, whey protein concentrate and lactose) are mainly sold as ingredients for the food industry.[7] *Per capita* consumption is therefore a less relevant measure, and Table 3.7 shows the absolute consumption of selected dairy ingredients in the EU.

Does the segmentation of Irish home demand for dairy products provide us with a possible explanation for the competitiveness of certain sectors in the Irish dairy industry? Table 3.8 shows Ireland's export position *vis-à-vis* other European countries. Ireland's share in total EU-10 exports indicate that Ireland has a competitive advantage in the production of butter (13.9%), cheese (7.9%), skimmed milk powder (13.7%) and casein (26.3%). This does not correspond with the structure of home demand. As we have seen, consumption of butter and cheese are low. Irish consumption of skimmed milk powder is also very low and casein is not a very important segment. Ireland has a weak position in the export markets for yoghurt (0.6%) and cream (0.3%). This corresponds with low *per capita* home demand for both products. On the basis of these figures one could conclude that the lack of home demand for yoghurt and cream led to a 'neglect' of these segments and was therefore detrimental for the competitiveness of Irish producers.

Table 3.5 Output flows of the Irish milk and dairy products industry, 1990

Customers for dairy products	Output flows million pounds	Share of total output (%)	Share of total sales in Ireland (%)
Agriculture/forestry/fishing	53	3	7
Milk and dairy products	254	12	36
Other food products	38	2	5
Beverages	13	1	2
Personal consumption	349	17	49
Services	7	0	1
Total sales in Ireland	714	35	100
Merchandise exports	921	45	
Invisible exports	13	1	
Changes in stocks	420	20	
Total output	2068	100[a]	

[a]Owing to rounding, items do not add to total.
Source: ESRI unpublished data.

6. The large increase in stocks during 1990 represents product sold into intervention. Most of these stocks will eventually be exported.
7. Skimmed milk powder and whey powder are also extensively used in the animal feed industry.

However, the findings regarding butter, cheese, skimmed milk powder and casein seem to contradict Porter's theory. This apparent conflict can be explained by a number of industry characteristics.

Until now we have disregarded a very important 'customer' for dairy products. The product mix and export portfolio of Irish companies are strongly related to the CAP support system of the EU. On the basis of export figures one might argue that Ireland has a competitive advantage in the production of butter, skimmed milk powder and casein. However, these markets are very much support/subsidy driven. The export portfolio is therefore not a reflection of 'real' demand conditions. Strong export segments cannot automatically be regarded as competitive.

There are additional reasons to believe that the competitiveness of some dairy products is not much affected by the structure of home demand. This

Table 3.6 Home demand (kg/capita) for selected dairy products in the EU, 1993[a]

Country	Liquid milk	Cream	Yoghurt	Condensed milk	Cheese	Butter	Ice cream
Germany	70.3	1.8	11.3	5.2	18.5	6.8	2.9
France	76.7	1.0	17.3	0.7	22.8	6.8	5.4
Italy	75.0	0.3	5.0	0.1	20.1	1.8	0.5
Netherlands	84.1	0.8	20.7	7.4	15.8	3.3	n.a
Belgium	65.4	1.5	5.7	1.1	19.8	6.9	3.3
Luxembourg	81.5	2.6	7.0	1.4	16.5	5.8	6.3
United Kingdom	114.9	1.1	4.6	2.6	8.3	3.5	0.3
Ireland	186.3	0.8	3.7	n.a.	5.7	3.9	n.a.
Denmark	114.7	2.9	8.3	0.0	15.4	4.1	4.9
Total nine	84.4	1.2	10.2	2.7	17.4	4.9	n.a

[a] Data for 1993 or the last year available.
Source: Milk Marketing Board, 1994.

Table 3.7 EU consumption of selected dairy ingredients (thousands of tonnes), 1993

Country	Whole milk powder	Skimmed and buttermilk powder	Whey powder	Casein/ caseinate
Germany	54	51	5	21
France	45	375	241	17
Italy	27	146	47	16
Netherlands	10	251	615	36
Bel./Lux.	17	5	n.a	4
United Kingdom	19	74	2	8
Ireland	15[b]	18	13	6
Denmark	9	11	23	1
Greece	6	8	3	2
Total 10[a]	202	939	949	108
Ireland/total 10	7%	2%	1%	5%

[a] Owing to rounding items do not always add to total.
[b] Cunningham and Pitts, 1995, p. 36.
Source: Milk Marketing Board, 1994.

holds for most of the commodity-type products at the end of the product life cycle (e.g. butter, skimmed milk powder, whole milk powder, etc.). There have been no substantial product or process innovations in these segments for some time. The technologies required to produce these products are readily available everywhere. Production decisions and sales are more dependent on fluctuations in the price that a product makes on the (spot) market than on the technological base and marketing capability of individual firms.[8] Demand for these products in one nation is unlikely to give a firm competitive advantage over its competitors abroad.

This does not imply that demand conditions have never been important for these products. However, the initial product development took place long ago and one has to go far back to trace a possible positive relationship between home demand conditions and competitiveness. The advantages of home demand have long been neutralized. A brief historical review of the Irish dairy industry does not provide much support for the idea that home demand contributed to the competitive advantage for the production of these products (Foley, 1993).

High butter consumption in Ireland might have contributed to a strong export position, although from an early stage production was strongly influenced by UK demand.[9] The first drying facilities were in place before the First World War and some of the skimmed milk powder was sold to industrial users. However, most of the by-product of butter manufacturing, skimmed milk, was returned back to the farmer for animal feed. The surge in skimmed milk powder production in the 1960s was partly instigated by the fact that farmers were no longer willing to have the skimmed milk returned to their farms. The strong market position of casein exporters does not seem to be a result of a strong home demand either. Although casein production started as early as 1932, the real development of the industry was instigated by strong North American demand in the 1970s.

The influence of home demand on competitive advantage was further investigated by asking managers whether experience in the Irish market regarding certain products was important for the companies' performance in international markets. The only obvious case is that of dairy spreads. Mixed fat spreads, a blend of dairy fat and vegetable oil, were first launched in 1969 in Sweden, as a response to the increasing market share of margarine. Ireland and the UK developed as one of the first new markets for spreads. When spreads were first launched in the Irish market they quickly gained considerable market share. Spreads constitute 42% of the Irish market for spreadable fats (butter, margarine and spreads) which makes the spreads

8. For example, this is illustrated by the 1995 shift in production from cheddar cheese to butter as a result of a better return for butter in third countries.
9. Before the introduction of dairy spreads in 1984 the per capita consumption of butter in Ireland was among the highest in Europe. (Pitts, 1996 forthcoming)

Table 3.8 Exports (thousands of tonnes) of selected dairy products by EU member states, 1993

Country	Liquid milk	Cream	Yoghurt	Con-densed milk	Cheese	Butter	Whole and semi-skimmed milk powder	Skimmed and butter milk powder	Whey powder	Casein	Ice cream
Germany	1657	113	126	293	349	72	98	426	277	30	18
France	770	63	69	56	414	98	231	69	284	30	35
Italy	7	1	0	1	109	21	0	0	9	0	4
Netherlands	81	45	17	308	463	258	236	95	128	8	7
Belgium/Lux.	800	22	93	45	100	153	106	106	94	2	45
UK	94	71	2	52	57	52	57	57	52	1	3
Ireland	57	1	2	3	153	114	41	122	46	31	1
Denmark	22	13	5	0	259	47	95	13	11	15	28
Greece	0	0	5	1	13	0	0	0	0	0	5
Total 10	3489	329	319	259	1917	816	863	888	901	118	146
Ireland/Total 10	1.6	0.3	0.6	1.2	7.9	13.9	4.8	13.7	5.1	26.3	19.2

aIncluding butter equivalent of butteroil.
Source: Milk Marketing Board, 1994; ZMP, 1995.

segment in Ireland one of the strongest in the EU. Per capita consumption of spreads is among the highest in Europe (Pitts, 1996).

All the managers who were exporting spreads were of the opinion that the early demand in the Irish market contributed to their competitiveness in international markets. One manager stated that they were able to offer foreign buyers any number of variations, something that had only come about because of what they had to do in the Irish market.

This is, however, the only example of a positive impact of home demand for a dairy product. Managers tended to stress the negative aspects of the home market, particularly in relation to yoghurts, chilled desserts and non-cheddar cheese. The *per capita* consumption of yoghurt, desserts and non-cheddar cheese in Ireland is among the lowest in Europe. Compared to the more mature markets of France and the Netherlands, Irish demand for these products developed relatively late. Processors in some other countries, for example France, experienced more favourable home demand conditions that might have led to an early competitive advantage.

The early development of strong competitors abroad probably contributed to the weak position of Irish producers in these areas. However, most of the managers stressed the small size of the Irish market as a more important factor. Branding is very important for success in products like yoghurts and desserts and it is notable that most of the companies did not even consider getting into these products. Establishing a brand requires substantial economies of scale because of the costs involved. As a result of transport cost and perishability of these products, preferential access to a large home market can therefore be an important advantage. This view was supported by some managers who stated that they would have developed these segments if the home market had been bigger.

3.4.2 Characteristics of customers in the Irish market

According to Porter (1990), a nation's firms gain competitive advantage if domestic buyers are among the world's most sophisticated and demanding buyers for the product, especially if their demand anticipates demand in other nations. Such buyers provide a window into the most advanced buyer needs.

The European consumer food market is affected by a range of social, economic and demographic trends, which have resulted in a rising demand for convenience products. A greater concern for health and environmental issues is responsible for a declining demand for full-fat products, while quality, hygiene and environment-friendly production processes have become important product attributes. Most of these changes reflect greater affluence and are occurring in all European countries, including Ireland (see Chapter 2).

In relation to consumption of low fat products, the Irish customer is a late mover. Table 3.9 shows that there has been a shift from whole milk to semi-skimmed and skimmed milk consumption in the EU. The Irish consumer has definitely not acted as an 'anticipatory buyer' in this regard. In 1993, half of the milk consumption in the EU was consumed as whole milk, down from 60% in 1988. The figure for Ireland was 92% in 1988 and has hardly come down since. This negative example of anticipatory buyer needs has probably not had a big impact on the competitiveness of processors in Ireland, since the production of low fat milk and full fat milk requires basically the same technology.

In relation to the aspects of service, quality and hygiene, we hypothesized that the sophistication of two local customer groups contributed to the competitiveness of the Irish dairy industry. We deal first with retailers. The role of retailers, and retail chains in particular, is shifting from a mere distribution channel for the dairy industry towards a customer for both branded and own-label products. Second, the role of multinational food and beverage companies located in Ireland is analysed because they are perceived as sophisticated customers for industrial dairy products.

Retailers. According to Porter, Irish milk processors could gain competitive advantage if domestic retail chains were among the world's most demanding buyers. Because of changes in consumer demand and technology the retail environment is changing, with major repercussions for the retailer–manufacturer relationship. In many countries the superstore has become an important retail format, as a result of price competition and changing consumer tastes creating a demand for wider product ranges (Collins, 1996). The introduction of information technology has proved important for reducing operating costs and facilitating the centralization of decision making, which has increased the buying power of the retailers. The buying power of the multiples has been further enhanced by the proliferation of private-label products.

Table 3.10 shows the concentration of the retail market in Europe. It is most concentrated in Scandinavia, where the top three retailers in each country control more than three-quarters of the market. Outside Scandinavia the market is most concentrated in Ireland, Belgium, Austria, the Netherlands, Germany and the UK. With 65% of the retail market controlled by the top three players (Power Supermarkets, Dunnes Stores and Musgrave-controlled retailers) the Irish retail sector is among the most concentrated in Europe. This higher level of concentration, helped by the fragmented nature of the Irish processing sector, has increased the buying power of Irish retailers. Kamann and Strijker (1995) suggest that the rela-

Table 3.9 Liquid milk sales by type (%) in the EU, 1988 and 1993

Country	Whole milk		Semi-skimmed milk		Skimmed and buttermilk	
	1988	1993	1988	1993	1988	1993
Total Nine[a]	60%	50%	32%	46%	8%	7%
Ireland	92%	91%	0%	0%	8%	9%

[a] Germany, France, Italy, Netherlands, Belgium, Luxembourg, UK, Ireland, Denmark.
Source: Milk Marketing Board, 1994

Table 3.10 European food retail structure, 1994

Country	Market share of top 3 retailers (%)	Penetration of private label (% of all sales)
Austria	56	21
Belgium	58	20
Denmark	77	15
Finland	80	5
France	38	21
Germany	46	25
Greece	17	9
Ireland	65[a]	27
Italy	11	8
Netherlands	47	16
Norway	86	7
Portugal	41	7
Spain	20	8
Sweden	95	10
UK	43	35

[a] This figure for Ireland (January 1995), presented by Goodbody, is considerably higher than that presented by ABN.AMRO because all the Power Supermarkets store chains are counted as one entity, as are the Musgrave controlled retailers.
Source: ABN.AMRO, 1996; Goodbody, 1995.

tively high producer concentration in Belgium and the Netherlands offsets the power of retailers, resulting in a relatively lower share of private-label products. Using the same logic one could argue that the high penetration of private-label products in Ireland (27%) and the UK (35%) could be a reflection of the strong bargaining power of the retail sector in these countries.

There are two aspects to the possible effects of these developments for the processing industry: cost efficiency and standards. Porter believes that large powerful chains can be a major force in constantly pressurizing manufacturers to reduce prices and thus costs, which leads to a more competitive position in international markets. He also argues that demanding buyers pressure local firms to meet high standards in terms of product quality, features and service.

Collins (1996) finds support for this in the UK. To minimize product failure, retailers demand stringent product specifications and access to the production facilities of the dairies. Plants are vetted, production runs are monitored and quality measured. Retailers are becoming more demanding in terms of product quality and innovation, logistics and relationship management. The relationship in its own right could lead to competitiveness. One of the direct implications of the involvement of retailers in private label production in the UK has been an increase in boundary personnel contact, e.g. between retail technologists and production managers or between retail buyers and key account managers. The exchange in and constant interplay between personnel assist the reduction in interfirm conflict and promote joint problem solving (Collins, 1996).

In their annual reports Irish dairy companies stress the importance of customer relationships, but in the interviews the responses of the managers diverged. Managers of the three big companies all stated that they attach great value to the development of interaction with retailers, a finding supported by an earlier survey of the Irish dairy processing industry (Varley, 1991). Companies have formed long-term relationships with local retailers, with key relationships evolving between the key accounts person and the customer. These relationships have developed beyond price negotiation and involve sharing of information and monitoring of service and delivery performance. Some retailers share information from barcode analysis. They communicate their ideas and work closely with processors in product development. However, all three companies stressed the central importance of costs. The relationship is important, 'but if somebody comes up with a better price the relationship is finished'.

The views of the managers of the smaller companies were less positive. None of them attached great value to the development of interaction with retailers. Three of the smaller companies did not supply the Irish retail maket. Another used agents for the distribution of its products on the Irish market. Although the remaining small companies dealt directly with local retailers and multiples, the interaction concerned the negotiation of prices only. Local retailers were considered as ruthless. The negotiations were 'cut-throat', involving no specification other than price. The companies experienced no loyalty from the retailers or multiples. One manager found more loyalty with the UK multiples which gave him more opportunity to develop. The situation is similar in the Dutch dairy industry where the demands of powerful retailers mainly relate to packaging and price rather than products (Jacobs et al., 1990).

The question remains whether or not the Irish dairy processors gain competitive advantage from their interaction with retailers or multiples in Ireland. There is some evidence of an effect on standards among the bigger companies. As described earlier, these companies all foster beneficial relations with the local retailers. Supplying local retailers had been a 'useful experience' when entering the UK retail market. One manager even stated that they were able to transfer some of the Irish experience into the UK. There seems to be more evidence of an effect on cost-efficiency. Almost all interviewees stated that the relations are principally cost based – 'the lowest price wins'. There was, however, no support for the idea that this was helpful in establishing a position abroad. On the other hand, one does not expect individual managers to be positive about the effects of an uncomfortable price-squeeze. In some cases it is probably contributing to a competitive advantage in international markets.

Foreign multinational enterprises.   In spite of the modest local market for food ingredients we hypothesized that the presence of multinational enterprises in

Ireland contributes to the competitiveness of the Irish processing industry. Large industrial buyers in Europe are becoming more and more demanding on issues of hygiene and process control (Kamann and Strijker, 1995). The findings of a study of the European dairy ingredients industry point to greater supplier participation in the drive towards innovation (PA Consulting Group, 1996). Suppliers are subjected to a rigorous screening process for inclusion on the approved supplier list and are assessed on quality, price, delivery and the capacity to live up to product specifications. Multinationals try to bind suppliers into long term relationships. A strong trend towards centralized purchasing and global sourcing leads to the added advantage of supplying several production facilities of the same multinational enterprise.

Foreign multinational enterprises always made up an important part of the local market for food ingredients, and a historical perspective of the Irish dairy industry provides some evidence of their contribution to innovation (Foley, 1993). From the early 1940s until the early 1970s foreign multinationals such as Rowntree Mackintosh were major local customers, buying chocolate crumb from indigenous processors. The early 1970s saw the development of a strong infant food manufacturing sector dominated by subsidiaries of large multinational enterprises such as Abbott, Wyeth and Cow and Gate (later acquired by Nutricia). Initially, these companies used partly demineralized fluid ingredients, including skim milk and whey, but the trend has been towards the use of customized dry ingredients. The Borden Company, which started production in 1961, has always been a major customer for whole milk powder. Their requirement for powder with the same physical properties that they had been accustomed to in the USA led to the installation of new dryers in Irish processing plants. Finally, foreign multinationals have also invested in the Irish beverage industry (e.g. the multinational Gilbey Group) buying cream and alcohol from Irish dairy processors.

Presently, foreign multinationals in Ireland are responsible for only a small part of the turnover of the processors interviewed. Two processors had no contact with foreign multinationals in Ireland at all, and five other processors sold them only a modest amount. However only one of the managers supplying to multinationals in Ireland did not support the view that supplying local multinationals had an impact on competitiveness and three managers believed that it had a significant impact. Most managers perceived a learning effect from dealing with the multinationals in Ireland. They learned about auditing standards, a codified approach to quality standards and purchasing systems. Multinationals were considered to be a step ahead in this respect. 'They were always critical of something which helped us internally.' In some cases the relationship resulted in process developments such as system improvements and improvements in plant layout which led to more consistency in product.

These findings seem to be in conflict with Porter's theory. According to Porter (1990), firms are better able to perceive, understand, and act on buyer needs in their home market. Understanding needs requires access to buyers,

open communication between them and a firm's top technical and managerial personnel, and an intuitive grasp of buyers' circumstances. Although Porter
admits that subsidiaries of foreign multinationals can play the role of sophisticated and demanding buyers, open communication is extremely difficult to
achieve with foreign buyers because of the distance from headquarters.

The multinationals located in Ireland seem to have played a positive role.
However, not all managers were of the opinion that this was an effect of easy
access and open communication between processors and local multinationals. In the first place, some companies had benefitted from selling to the
local multinationals without developing a substantial relationship with them.
Apart from, that it is debatable whether the relationships that have developed are a result of the Irish location of these multinationals. Only one manager, of a smaller dairy company, stated that proximity was important.

> Proximity of a few customers in Ireland lends itself to working in a special way
> with them. We obviously also have long-term relations with customers abroad, but
> we are doing things with them (the multinationals located in Ireland) that you
> could not do abroad. You need to be on the same island.

However, the other companies had been able to foster the same relations
with customers abroad. Most of the companies, including the smaller ones,
meet their customers face to face on a regular basis, either in Ireland or
abroad. 'We send people out there, travel is no impediment any more.' In the
case of the smaller processors the relationship with foreign customers is
often facilitated by the IDB, which brings in the customers from abroad for
factory visits and to discuss product changes. It also facilitates visits by Irish
processors to customers abroad. There is no evidence of variation in customer feedback due to geographical distance. One manager of a larger
processor even stated that he had probably been more successful in developing advanced relations, and probably developed more sustainable value
added products, with customers outside Ireland.

3.4.3   The importance of home demand conditions

Porter's concept of segmentation proved to have some relevance in the
analysis of competitiveness, although less so for commodity-type products at
the end of the product life cycle. The evidence that home demand for individual dairy products had a positive influence on competitiveness is limited
to the case of dairy spreads. We found more support for the idea that the
lack of home demand for individual dairy products like yoghurt, desserts and
non-cheddar cheese had a negative influence on the competitiveness of the
Irish dairy industry. Managers stated that they would be producing these
products if the Irish home market had been bigger.

The competitiveness of the Irish dairy processing industry seems to be
more positively influenced by the characteristics of customers located in the
Irish market, in particular the retailers and the multinational food and bev-

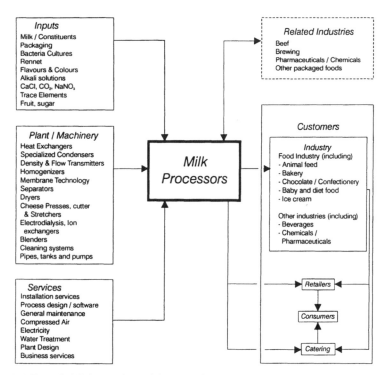

Figure 3.4 Extended linkages chart of the Irish dairy processing industry: potential.

erage companies. The retail sector in Ireland is one of the most concentrated in Europe and is a sophisticated and demanding buyer. Their requirements in terms of product quality, features and service have upgraded the standards of the larger Irish dairy processors to some extent. However, a more important implication of the increased bargaining power of the retailers has been the constant pressure on the processors to reduce prices, and the consequent effects on cost efficiency.

The overseas multinationals located in Ireland constitute a second group of demanding and sophisticated buyers for the Irish dairy processing industry. Although the multinationals are responsible for only a small part of the turnover of most processors, they appear to have contributed to the competitiveness of the dairy industry. Most processors learn from the standards and systems employed by the multinationals in Ireland, an experience that helps them in international markets.

3.5 Related and supporting industries

In this section we analyse the importance of the third broad determinant of national advantage, the presence of supplier industries or related industries which are internationally competitive. The most important inputs of the dairy processing industry are outlined in Figure 3.4. Milk and milk constituents are by far the most strategically important inputs. The only other input of some strategic importance is packaging. Plant and machinery includes both very sophisticated equipment, such as ion exchangers and specialized condensers, and more basic items such as tanks, pipes and pumps. Finally, installation work and process design and software are the most strategically important services.

Related industries are, in principle, country specific. Therefore, Figure 3.4 does not show all the industries that could possibly be related to the dairy processing industry. Instead we present the industries that, on the basis of interviews with key informants and information from secondary literature, were hypothesized to be related to the Irish dairy industry. In section 3.5.1 we examine the actual importance of these potentially related industries for the competitiveness of the Irish dairy industry

3.5.1   Dairy farming

The main input required by the dairy processing industry is the milk supplied by farmers. For this reason, the quality of the milk, the milk price paid to the farmer and the farmer–processor relationship have received considerable attention in many previous studies. However, it has proved difficult to determine the exact influence of these issues on the competitiveness of the dairy processing industry, partly because of the complex relationship between the issues.

In order to determine the impact on competitiveness we need to understand that the farmer–processor relationship is complicated by the fact that the farmer is both supplier to, and the main shareholder in, the processors. This link between the shareholder and the supplier is an important issue for the competitiveness of the Irish dairy processing industry and we will return to this issue in the next section on firm strategy, structure and rivalry. Here, we concentrate on the issue of milk as an input to the processing industry.

In 1994, the average milk producer price in Ireland and Spain was the lowest of all European member states (ABN.AMRO, 1996). *Prima facie*, a relatively low milk price would suggest a competitive advantage for processors in Ireland. However, when comparing milk prices between countries it is necessary to control for differences in bonus systems, seasonal payments, VAT, currency translations and the location of measurement (at farm or at dairy). Furthermore, in assessing the advantage of a low milk price one must account for differences in milk composition, collection costs and seasonality of supply.

In comparison to the milk processed by Ireland's main European competitors, Irish milk has a low fat and protein content (ZMP, 1994). Irish processors are also confronted with higher milk assembly costs than competing European countries (Forbairt, 1995). The low milk producer price is mainly a reflection of the low fat content of Irish milk (ABN.AMRO, 1996).

However, Table 3.11 shows that, even when we control for these variables, the milk price paid by Irish processors is still low compared to their main European competitors.

The figures in Table 3.11 do not take account of an important characteristic of the Irish milk delivery pattern: the highly seasonal nature of milk supply. An even pattern of milk supply is generally considered an advantage for milk processors, but it usually increases the costs to the farmer. Irish milk is produced in a pronounced seasonal manner. Keane (1995) shows that seasonality accounts for many of the scale inefficiencies at cheese processing plant level. In our research a manager of one of the larger companies indicated that, due to seasonality, a major plant is operating well below full capacity. Clearly, the seasonal delivery pattern partly offsets the advantages of a low price.

So, the advantage of a relatively low (standardized) price for milk realized in Ireland is partly counterbalanced by a highly seasonal supply. It has proved difficult to establish the exact extent to which the advantages of a low milk price outweigh the disadvantage of seasonality (Keane, 1995). The contribution depends on several factors which are difficult to control for, including the (feasible) product portfolios of the processors and the pending changes in the CAP.

3.5.2   Other supplier industries

The second most important group of suppliers are the competing dairy processors themselves. Irish dairy processors receive an important part of their raw material inputs from other dairy processors. There is a significant amount of trade in raw milk between neighbouring processors, both in response to market forces and in order to create production efficiency.

Some of this trading involves *ad hoc* arrangements, depending on the economic conditions of the day, but most trading involves long-term (though rather informal) arrangements. These latter arrangements contribute to the efficiency of the Irish processing industry. A good example is the 'winter-working' arrangements involving the transfer of milk to a neighbouring processor for a few weeks while the plant is closed for maintenance. Some

Table 3.11 Milk prices (delivered to dairy) as a percentage of the target price, 1989–1995

	1986	1987	1988	1989	1990	1991	1992	1993	1994	1995
Belgium	90.8	90.6	95.5	100.4	91.9	89.4	91.5	92.9	89.0	91.0
Denmark	101.5	98.6	107.2	109.2	108.5	107.8	105.3	101.5	101.0	99.0
France	89.7	91.4	94.0	93.8	93.3	91.5	93.0	94.2	95.5	95.0
Germany	94.5	92.6	98.4	107.3	101.2	99.0	100.1	98.4	96.5	93.0
Ireland	90.8	90.6	95.5	101.1	89.7	86.5	92.9	90.6	90.0	96.0
Netherlands	93.8	94.3	100.8	104.2	96.7	98.9	101.6	100.7	99.0	98.0
UK	89.8	89.8	94.1	95.2	93.2	92.6	90.2	86.7	94.0	98.0

Source: ABN.AMRO, 1996.

milk trading is instigated by changing milk characteristics. Processors buy or sell milk depending on the seasonal milk characteristics most suitable to their production.

There is also substantial trade in milk by-products such as skimmed milk, butter milk, whey and casein. The individual processors try to optimize the use of their production capacity by buying the milk components they need for their own production and selling by-products to other processors. This trade between the Irish dairy processors of course leads to greater efficiency, but not to a competitive advantage for the Irish industry. One could argue that the trade merely compensates for an inefficient industrial structure (many relatively small processors) which leads to trade between companies in search of a sufficient and suitable supply of raw material. The larger Irish dairy processing companies are less dependent on trade for their raw materials since they are able to transfer milk and milk by-products between their plants.

Apart from milk and milk by-products, the number of inputs of strategic importance for the dairy processing industry is limited. The most important non-farm supporting industries are packaging, equipment manufacturing, installers (engineering companies), and the software industry. In some countries dairy processing benefits from the presence of competitive supporting industries. In his own study, Porter (1990) presents the example of the Danish dairy industry, which is well serviced by local companies producing food processing equipment and refrigerating equipment. Jacobs *et al.* (1990) point to the importance of the local dairy equipment (e.g. Stork) and installation (e.g. Hovap) industries for the Dutch dairy sector.

The supporting industries in Ireland are less developed than in Denmark or the Netherlands. Nearly all companies source at least part of their packaging requirements from companies located in Ireland. One packaging supplier (Smurfit) is a global competitor. The impact on competitive advantage appears to be limited. Although one manager stated that the local suppliers are cost efficient, two managers rated them as expensive (one of these two managers described the situation until recently as 'a near cartel situation' which had to be broken by imports), and the other managers took a neutral stance. The quality of the packaging materials produced in Ireland is generally perceived as satisfactory. We found some evidence of interaction between processors and packaging suppliers and many dairy co-operatives share ownership of a co-operative packaging company. However, in most cases the interaction is limited to the sharing of information concerning market developments. We encountered no situations where interaction led to upgrading or innovation.

Nearly all the required machinery is produced abroad by the major German and Danish companies (Niro, Alfa-Laval, Tuchenhagen, APV), who in many cases are also responsible for the yearly overhaul. Some Irish companies are involved in the manufacture of basic equipment such as tanks, pipes and pumps. Installation work, process design, and maintenance is carried out by both foreign and local companies. A number of indigenous companies are concentrating on plant design and project management. Three managers stated that they employ Irish companies. One manager saw Ireland as 'well serviced' by price efficient engineering companies.

The market for plant software is again dominated by the large international companies, but some work is carried out by indigenous companies. Four processors use local software companies. These companies are mostly hired for the customization of process control systems that come with foreign produced equipment and plants. At least one of these local software companies, Ogenek, has been very successful in exporting its dairy software systems. Although two managers stated that the interaction with local engineering and software companies has developed into more than a simple transaction, the interaction did not appear to involve more than some information sharing.

The general impression is that the supporting industries in Ireland do not make a significant contribution towards the competitive advantage of the dairy processing industry. Almost no local suppliers have developed into global competitors, Smurfit being the exception. Most can be described as captive industries, mainly supplying the local market. Furthermore, relationships with local suppliers appear not to be that much different from the relationships with suppliers abroad: 'We have the same level of interaction with suppliers abroad. Abroad is no further than Ireland' (interview respondent). There is no evidence of substantial co-ordination or joint development with local suppliers. In fact, the only examples of joint development and early access to new technologies involve suppliers located abroad. For example, a UK subsidiary of an Irish dairy company gained early and exclusive access to a new heat seal development because of its relationship with a French packaging company.

3.5.3 Related industries

Porter (1990) describes related industries as those in which firms can co-ordinate or share activities in the value chain or those which involve products that are complementary. Sharing of activities can occur in technology development, manufacturing, distribution, marketing or servicing. The presence of competitive related industries could enhance the competitiveness of the dairy industry. Porter provides the example of Denmark, where the dairy industry benefited from the presence of strong insulin and enzyme producers with whom it shared common technologies and skilled workers. The examples in Ireland are less convincing. The dairy industry has links with related industries through diversifications into other food areas. Apart from that managers gave examples of less direct links, notably with the chemicals/pharmaceutical and brewing industry. These links with other industries do not appear to have a strong impact on the competitive advantage of the dairy processing industry.

An obvious case is the dairy industry's active manufacturing involvement in related food industries. A number of Irish dairy companies are operating in areas as diverse as meat, orange juice and mineral water. This diversification is not entirely new to the dairy industry. Irish co-operatives have always been involved in the processing and sale of farm products combined with the manufacture of farm inputs. The different activities are very much interlinked. Farmers buy inputs including animal feed from the co-operative and sell their milk to the dairy division of the same co-operative.

There was little support among the managers for the idea that the dairy industry benefits from its links with the meat industry. It is possible that the dairy industry and the meat industry employ workers with similar training, but the dairy industry is believed to have its own 'high' standards. It is possible that the two sectors benefit from each other's knowledge regarding (vacuum) packaging, but most managers claimed that the meat sector gains more from the presence of a strong dairy sector than the other way around.

A more recent development is the diversification into the processing of non-farm raw materials, such as orange juice, mineral water and soup. These moves are facilitated by production and distribution synergies. The processing and bottling of milk, orange juice and mineral water require similar technologies and the same distribution network can be utilized for the marketing of the products. Soup packaging also builds on the available packaging expertise of the dairy industry. The number of diversifications in these areas is small and the impact on competitiveness of the dairy division, questionable. According to one manager the main advantage is increased group turnover (and profits) plus adding a product to the 'basket' on offer to the multiples.

Apart from their direct involvement in related production, the managers also gave examples of other links with related industries. Most cases involved the pharmaceutical and brewing industries. Pharmaceutical, brewing and dairy companies use similar technologies. We encountered some concrete examples of shared technology. One processor 'looked at the technology and control system of a local brewing company'. Another processor was able to 'help one of the pharmaceutical companies with holding quality through the production process'. A number of managers stated that they employed technical staff with experience in the pharmaceutical industry.

The full extent of beneficial relations between industries is probably greater than the individual players might perceive. This is because many of the benefits are derived through indirect links, for example, through suppliers. This appears to be the case in the dairy industry also. One processor benefited from an Irish engineering company using technical expertise gained in the pharmaceutical industry. The full name of one of the engineering companies, Brewery, Chemical and Dairy Ltd (BCD), is illustrative in this regard. Similar transfers possibly occur via the software industry given the fact that a noticeable number of software companies has customers in different processing industries such as dairy, drinks, chemicals, pharmaceuticals and food.

This transfer of knowledge through suppliers is enhanced because of new company formation by employees of the different processing industries in Ireland. A noticeable example is the case of Golden Vale's engineering division, wound up in the early 1980s. This led to the formation of a new engineering company, with the result that some of the engineering expertise of the dairy industry became accessible to the other processing industries in Ireland. In a similar fashion, some of the processing technology of the brewing industry was 'unlocked' when a number of engineers employed by Guinness left the company to set up their own business and started servicing the dairy industry.

3.5.4   The importance of related and supporting industries

The most important suppliers of the dairy processing industry are the farmers. We examined the contribution of the milk price towards the competitive advantage of the Irish dairy processing industry. For a number of reasons it is difficult to determine the exact contribution. First, although a low milk price could be indicative of efficient farming, it could also be a result of inefficient processing. Second, milk price comparisons fail to take account of differences in the governance structure of the processing industry. These differences affect the way farmers are 'rewarded', i.e. through a maximization of milk prices or through the optimization of corporate profits which accrue to (farmer-) shareholders. Although inconclusive, the data suggest that Irish farmers, compared to farmers in most other European countries, are competitive producers. But, as we have seen, Irish processors have to deal with a highly seasonal supply which partly offsets the advantage of a low price for the main milk constituents.

Competing dairy processors, supplying milk or milk by-products for further processing, constitute the second most important source of raw materials. Although the trade between the individual processors might lead to greater efficiency in the Irish processing industry, it can be argued that it only represents an attempt to compensate for an inefficient industrial structure.

Although there are some examples of links with related industries, again, the general impression is that the impact on the competitiveness of the Irish dairy industry is limited. The diversifications into related food industries had little impact on the competitiveness of the dairy divisions. The links with the pharmaceutical and brewing industries appear be more interesting and suggest potential for further development. Some suppliers appear to have benefited from the presence of competitive processing industries located in Ireland, and we found some cases where this led to an indirect transfer of knowledge between related industries. However, as yet, most links seem to be incidental rather than part of a constructive network of related industries.

3.6 Firm strategy, structure and rivalry

3.6.1   Strategy and structure

The Irish co-operative movement celebrated its centenary in 1994. Co-operatives have dominated the Irish dairy industry and this is likely to be positive for commitment in a Porterian sense. However, over the last decade four of the five major players have adapted their structures to various forms of a hybrid co-op/plc structure. Harte (1995) argues that, while the expressed objective was to design a new funding mechanism, the solution had a much more far-reaching affect. It provided a means by which many of the problems associated with vertical ownership and with the co-operative structure could be overcome. It addressed the lack of incentives for staff, the investment portfolio problem and the restriction imposed by the horizons of farmer members (Harte, 1995).

The industry remains highly concentrated in the core business of dairying (Table 3.12). In particular, the smaller co-ops, together with Waterford and Golden Vale, are highly focused on dairying. The other three larger firms have reduced their dependence on dairying.

Table 3.13 shows the size of the larger Irish companies relative to the top companies in Europe. The companies are ranked on the basis of litres

Table 3.12 Turnover in dairying as percentage of group turnover

Company	1994
NCF	32
Mid West[a]	74
Shannonside	100
GV	88
Nenagh	67
Avonmore	50
Lakeland[b]	90
IDB	
Tipperary	90
Carbery	100
Waterford[c]	90
Kerry[d]	30
Dairygold	51

[a] Balance is trading and stores.
[b] Estimate.
[c] Figure includes dairy and consumer products.
[d] Not possible to determine, but 'guesstimate' of dairy derived sales is 30%.
Source: Derived from company interviews and annual reports.

Table 3.13 Ranking of Irish dairy enterprises on size and profitability criteria (1994 figures)

	Litres processed	Dairy sales	Dairy operating margin
Number 1	Besnier	Nestlé	Nutricia
Waterford	13	21	17
Avonmore	11	20	15
IDB	n/a	9	23
Golden Vale	20	23	n/a
Dairygold	19	25	n/a
Kerry	22	26	13
Selected foreign companies			
MD Foods	4	8	14
Dairy Crest	14	14	19
Friesland	10	7	18

Source: Compiled from information presented in ABN-AMRO (1996) and various annual reports.

processed, dairy sales and operating margin. It is clear that Irish companies are still relatively small in size, but the ranking of all the major Irish processing companies is improved when operating margins as opposed to sales data are used as the criterion.

The IDB plays a significant role in the Irish dairy industry. In this section we also discuss the position of the IDB in the strategies employed by the companies. The strategies of the dairy processing companies vary considerably, but it is possible to discern two broad patterns. Companies are categorized into two groups, on the basis of company interviews and annual reports (Table 3.14). The first group comprises 10 companies for whom the driving force is milk price. In this group there are a number of smaller co-ops whose pursuit of this strategy is supported by the IDB. Further, there are a number

Table 3.14 Strategic groupings in the Irish dairy processing industry ($n = 13$)

Companies	Characterization
Group A	Primary goal – milk price Produce commodities at lowest cost Utilization of regional resources important for some Maintain healthy financial base Adding value in various markets stated as an ambition
Group B	Primary goal – ROI Expansion orientated Market development and product development Strong international orientation.

Source: Derived from company interviews, industry interviews and annual reports.

of larger co-ops who have adopted a slightly longer term and broader perspective which is more market-driven. These companies certainly were, and to a large extent continue to be, facilitated by the export opportunities offered by the IDB. The extent to which these companies use the IDB varies. This second group comprises three large companies which are expansion orientated in order to achieve highest return on investment. Nonetheless, these companies are still heavily influenced by the 'milk price factor'. These companies vary in their use of the IDB, but it does provide an important outlet for their products.

In order to fully understand these strategies it is important to realize that the supplier is the shareholder whose primary goal has centred on the development of their core business, namely farming. In practice this means achieving a higher milk price. To date, Irish processors have been driven by this short term pressure and have focused on low cost production of commodities. Kamann and Strijker (1995) report the same phenomenon in the Netherlands and suggest that the goals of farmers contradict the long-term goals of the processing industry in modern times (p. 140). According to some of the interviewees in this research, the ownership structure and political importance of the milk price has meant that the industry has not re-invested sufficiently in the upgrading of facilities, a fact which will become apparent when market conditions become more difficult. However, an examination of Table 3.1 suggests that the Irish industry has achieved high growth in profitability. Thus we may conclude that the investment by industry has been to achieve greater efficiencies. However, recent levels of investment may not have been sufficient.

We now examine the extent to which the particular structure and the broad strategies adopted by companies have been influential in improving the competitive position of the Irish industry. The product, investment and research strategies as well as managerial motivations are examined.

Product strategies.   The Irish dairy processing industry in general is criticized for its overdependence on commodity-type products and lack of breadth in its product portfolio.[10] The product portfolio has a heavy emphasis on butter, with an allocation of 62% of Irish whole milk, on an output basis, versus only 17% to cheese in 1994, compared with 65% and 14% in 1991 (IDB, 1995, p. 3). These allocations have not changed significantly since EU membership, whereas Ireland's main dairy competitors have all reduced their reliance on butter.

To understand the product portfolio it is necessary to clearly understand the evaluative criteria used in the comparison. The term 'value added' is syn-

10. The reasons offered have included remoteness and a lack of home market, risk and financial considerations, management skills and investment decisions, farm production and structural problems, low R&D and marketing investment (see for example O'Sullivan, 1982; Keane, 1984; ICOS, 1987).

onymous in dairy industry reports with the term 'higher margin'. This confusion is not helpful, as it fails to recognize the source of the added value; in particular, it does not address whether the value is actively sought or passively acquired. An alternative measure developed by Nixon (1995) distinguishes between the active and passive adding of value in the dairy industry.

- Passive value added is outside the control of the firm. This could be due to relaxation of tariffs or movements in the commodity cycle (Nixon, 1995, p. 4).
- Active value added, occurs when a firm takes decisions to add value deliberately by differentiating its product from other competing products, through various marketing techniques.

The companies provided a confidential classification of the breakdown, by type of product, of company sales in 1995. The products were allocated into liquid milk, skim milk, skim milk derivatives (including casein and caseinates), whole milk powder and derivatives, butter and butter derivatives (including spreads and butteroil), cheddar and non-cheddar cheese, whey and whey derivatives and 'other'. The value of sales in 1995 under the headings of skim milk derivatives and butter derivatives and whey derivatives, along with some non-cheddar cheeses and other customized products, is used as an indication of 'active value added'. The results outlined in Table 3.15 indicate that Ireland has about a quarter of its products in the active

Table 3.15 Regional product portfolio of Irish dairy companies in 1995, expressed as a percentage of production by Irish companies in the respective regions[a]

Product categories	Product portfolio (Irish based production) (%)		Product portfolio of Irish dairy companies (All regions) (%)	
SM derivatives (mainly casein)	15		11.1	
WMP derivatives	1		0.6	
Butter derivatives	3.0		1.8	
Non-cheddar	4.7		10.8	
Whey derivatives	0.7		0.5	
Other	0.5		0.9	
Sub-total		24.9		25.7
Liquid milk[b]	22.6		31.8	
Basic SMP	15.5		10.5	
WMP	4.3		4.3	
Butter	25.3		15.1	
Cheddar	6.5		11.5	
Whey	0.8		1.0	
Sub-total	75		74.2	
Total		99.9		99.9

[a] The IDB is excluded from this analysis to avoid double counting and Kerry and Dairygold are not part of this analysis.
[b] Includes fresh products for Waterford.
Source: Derived from survey.

value added category. If casein is excluded, the proportion falls to about 10%. It is apparent that in many cases companies are increasing value added through the formation of close relationships with industrial buyers of food ingredients and cheese both in Ireland and abroad. Further a number of company strategies are similarly dominated by the need to 'get closer to customers'. Relationships have been highlighted in a number of reports (for example Varley, 1991; PA Consulting Group, 1995) as a key component of the industry's marketing strategy. In this context PA Consulting Group (1995) highlight a number of concerns in relation to Irish dairy companies and their approach to relationships indicating that there is scope for improvement. In our research it was apparent that a number of companies may be paying lip service to marketing ideas, for example, on relationships. One interviewee indicated that 'at the end of the day price is all that matters'.

Notwithstanding these reservations, this analysis concludes that the companies, although still heavily orientated to commodity-type products, are finding a middle ground between basic commodities and higher margin products by focusing on the development of relationships.

A further interesting finding from Table 3.15 is that the product portfolio of companies is more balanced than may have been apparent by the output figures presented above. Liquid milk is the main product while the allocation to non-cheddar cheese (mainly processed cheese) is 10.8% versus 11.5% for cheddar. However Irish based production still has a heavy emphasis on butter (25% of sales value in 1995). This suggests that a consequence of the international expansion of production by the three companies has been a more balanced portfolio. It is especially apparent that the reliance on butter is reduced.

Investment strategies.   A further concern is whether companies are carrying out the more active value added activities abroad; more generally, what is the nature of and impact from outward investment by three Irish dairy companies[11] on the Irish processing industry.

An examination of acquisitions in the dairy sector between January 1985 and April 1996 by companies based in various European countries shows that the French and British have been the most active, with Ireland ranked fourth behind the Netherlands (Table 3.16). Table 3.17 shows the net acquisitions,[12] and on this basis Ireland is ranked third behind France and Denmark demonstrating clearly that the Irish industry has been very outward-oriented in its strategic focus. Table 3.18 chronicles the acquisitions by all Irish dairy companies since 1987, showing that most of the acquisitions

11. Three companies from the sample have invested in dairy processing abroad. The two non-participating companies also have activities abroad. Kerry have concentrated their expansion abroad and Dairygold acquired their only foreign dairy processing facility (in Somerset, UK) when they bought a cheese manufacturer in 1994.
12. Net acquisitions is acquisitions in a region less sales in that region.

Table 3.16 Main acquisition activity in different EU countries, January 1985 to April 1993

Base of acquiring company	Home country	Other EU country	Rest of the world	Total
France	76	81	42	199
UK	98	24	24	146
Netherlands	20	16	32	62
Ireland	**21**	**24**	**11**	**52**
Italy	20	7	21	48
Germany	18	2	10	30
Spain	25	0	0	25
Denmark	7	8	2	17

Source: Perez *et al.*, 1994, p. 168; ABN AMRO, 1996, p. 20

Table 3.17 Net acquisition[a] activity in different EU countries, January 1985 to December 1993

	EU	Rest of world	Total
France	70	31	101
Ireland	14	10	24
Denmark	11	26	37
Netherlands	7	1	8
Total	102	68	170

[a]Net means acquisitions less sales in each region.
Source: Perez *et al.*, 1994, p. 170.

have been of UK and US companies, and to a lesser extent non-UK European-based companies.

Table 3.19 examines the product portfolio of three dairy companies with international production facilities. From Table 3.19 we can calculate that 53% of dairy sales by the three multinationals were of dairy products produced abroad. The biggest product category produced is liquid milk. There was little evidence among the three companies that there is a trend towards locating active value added activities or of producing higher margin products abroad.

The central issue of importance in this section is to understand the motivation and consequence of these investments for the Irish dairy processing industry. We asked the interviewees to discuss the reasons they had internationalized production and the benefits that had emerged. The companies retain close control of their subsidiaries which improves the chances of benefits flowing back to the home country but also diminishes the extent to which the subsidiaries are likely to integrate into the foreign diamond. The companies have invested in foreign markets in either production or distribution facilities, and the details of these acquisitions are contained in Table 3.18.

The need to acquire scale and access to raw materials was cited by all the respondents as a reason for investing abroad, and greater opportunities for this kind of structural adjustment were available abroad. Further, many of the investments incorporated a distribution network. This is a major benefit

Table 3.18 Major foreign dairy related acquisitions by Irish dairy companies

Irish company	Year	Acquisition company	Country	Business	Cost
Kerry	1987	Primas Fd Ing.	US	Food ingredients	n.a.
Kerry	1988	Beatreme	US	Food ingredients	US$120M
Avonmore	1988	Roy's Dairies	US	Dairy processing	US$9.5M
	1988	Glenmills Dairies	UK	Liq. milk	IR£0.5M
Avonmore	1989	St Falbo Cheese	US	Cheese mfg.	n.a.
	1989	Golden Dairies	UK	Mozzarella	n.a.
Waterford	1989	Heald Foods	UK	Milk/F. juice	IR£43.0
Golden Vale	1989	DPP	UK (NI)	Proc. cheese	IR£5.0M
	1989	Ceredigan	UK	Liq. milk	n.a.
	1989	Golden Cow	UK (NI)	Butter spreads	n.a.
Tipperary	1989	CPL Davione	France	Cheese	n.a.
Avonmore	1990	Birmingham Dairies	UK	Liq. milk	
	1990	Goodwins	UK	Liq. milk, cheese	GB£5.7M
	1990	Handsworth	UK	Liq. milk	GB£7.7M
Kerry	1990	Milac GmbH	Germany	Food ing.	n.a.
	1990	Semmons Taylor	UK	Food ing.	n.a.
Waterford	1990	Galloway West	US	Dairy products	IR£44.9
	1990	Western Cheese	UK	Cheese	n.a.
Golden Vale	1990	Bridgend	UK	Liq. milk	IR£3.6M
Waterford	1991	U.C. Dairies	UK	Milk	IR£40.8
Avonmore	1991	Caterpak	UK	Grated cheese	IR£0.4M
Avonmore	1992	Wards Cheese	US	Cheese mfg.	n.a.
	1992	Whitcroft Dairies	UK	Liq. milk	IR£4.4M
	1992	Wiltshire Dairies	UK	Liq. milk	n.a.
	1992	Hampshire Dairies	UK	Liq. milk	n.a.
	1992	Golden Foods	Belgium	Cheese mfg.	n.a.
	1992	Churchfield	UK	Liq. milk	GB£5.7
	1992	Parker	UK	Liq. milk	GB£4.9
	1992	Paszto Kft	Hungary	Liq. milk	n.a.
Golden Vale	1993	Leckpatrick	UK (NI)	Dairy pdts.	IR£22.2M
	1993	Vonk Food Holland	Netherlands	Processed cheese	
	1993	A/S Vejle	Denmark	Margarine	
Waterford	1993	Durham Dairies	UK	Liq. milk	IR£7.7M
	1993	Express (NI)	UK	Mozarella	n.a.
Avonmore	1993	Dairycrest	UK	Liq. milk	IR£21.6M
Waterford	1994	Greencroft Dairies	UK		n.a.
Kerry	1994	DCA	US[a]		US$402M
Dairygold	1994	Horlicks	UK	Cheese	n.a.
Waterford	1995	TCC	UK	Dairy Foods	IR£125.0

Source: Annual reports and other industry material.

as distribution is a critical success factor in international markets. The internationalization of production was also a response to the need to acquire geographical diversity to decrease or spread risk. Other benefits which arose because of the investment decisions included more effective development of relationships which are now perceived as the basis for future development. A further benefit from these investments is improved access to research and development facilities. Further all companies felt it allows them to deploy their resources more effectively, both capital and human. The companies did

Table 3.19 Percentage of group sales and absolute value, by product category, accounted for by production outside Ireland, in 1995 ($n = 3$ companies)

Product category	Production in Ireland (%)	Production outside Ireland (%)
Liquid milk	32.1	44
Basic SMP	14.8	4
SM derivatives (mainly casein)	15.2	6
WMP	3.1	4.4
WMP derivatives	0	0
Butter	21.1	1.7
Butter derivatives	4.4	0.3
Cheddar	4.4	18
Non-cheddar cheese	3.6	18.9
Whey	1.1	1.2
Whey derivatives	0.3	0.4
Other	0	1.4
	100	100
Total company dairy production (£m sales, in 1995)	868.7	1005.24

Source: Derived from survey.

feel that their physical presence may lead to some extra access to information on customers: the 'bodies on the ground' effect. Exposure to foreign competition had created extra pressure to innovate, but this is not due to the actual investment in the countries *per se*. There are other benefits and links, including knowledge concerning milk suppliers, human resource policies from the USA and benchmarking technology which the companies have been exposed to. On balance the companies felt that Ireland was a net exporter of knowledge. Finally, it was apparent that many of investments represented an opportunity to use skills in the area of cost reduction

These findings suggest that there are some learning benefits for the home base from the international investments. However, is this learning significant for the competitive advantage of the industry? Dunning (1992, p. 145) suggests that companies can have different forms of investments, namely, resource seeking, efficiency seeking, market seeking, or strategic asset seeking, and that each form of investment represents a different learning experience. It appears that the driving force behind the Irish dairy companies investments has been the need to acquire scale, not readily available in the home market. The emphasis has been on acquiring resources and access to markets. The investments do not seem to have the specific of objective of transferring assets back to the home base, nor is the attainment of efficiencies a dominant goal. However the net flow of knowledge is considered by most to be outwards rather than inwards. Further, the companies expressed the view that while many of the benefits are magnified by the actual direct international investment, many of the benefits would accrue from an international marketing presence.

In conclusion, it is clear the reasons offered for internationalization are mainly concerned with acquiring greater access to resources. In a sense this is line with Porter's hypothesis that companies will invest abroad in order to exploit and upgrade competitive advantage, created in the home base. Finally, while it seems reasonable that companies are 'tapping' into foreign diamonds, learning appears to be limited and is secondary to the primary goal of increasing scale.

R&D strategies.   A further indication of the orientation of Irish dairy companies is provided through an examination of their R&D strategies. This section outlines the levels and nature of R&D among Irish dairy companies.

Forfás[13] found that in 1993 the spending on R&D in the sector was 0.38% of sales, which is an increase relative to 1988 (0.24%). However, this is less than the food sector generally, which spends 0.5% of sales. In absolute terms the spending by the 13 companies amounted to IR£15.2 million in 1993, an increase of over 50% relative to 1988 (IR£7 million). However, in practice 95% is by the big five plus IDB. The industry is dependent on its own funds (84% versus 86% for all manufacturing and 78% for all food) for research, and in particular this is the case for the smaller co-ops. In 1993, 93% of research expenditure was on R&D conducted internally.

The structure of and attitude towards R&D among the companies varies considerably. Forfás (1995) find that among 'R&D performers' in the dairy industry 8 of 13 dairy companies had a 'formal R&D department'. Our findings (Table 3.20) confirm that most of the smaller companies have either a very informal approach to R&D or no internally conducted R&D. It is also possible to examine the form of R&D expenditure. Forfás provides estimates as outlined in Table 3.21. The data show that 44% of research in 1993 was for new products which is lower than the rate found in the industry in general (51%). Furthermore, our research suggests that research into new products in this industry is in areas new to the company rather than new to the market.

Table 3.20 R&D structures in Irish dairy processing industry

	No. of small companies	No. of large companies	Total
No R&D dept.	1	0	1
R&D informal	3	0	3
Head of R&D[a] (no full time staff on R&D)	1	1	2
R&D department	1	1	2
Integrated R&D department (integrated team approach)	1	2	3

[a]More looking for new ideas or opportunities than R&D.
Source: Survey.

13. Unpublished data; sample included 13 R&D performing companies.

Table 3.21 Breakdown R&D expenditure by objective of research

	%
Improving existing processes	18
Developing new processes	17
Improving existing products	21
Developing new products	44

Source: Forfas, unpublished data.

In conclusion the sector and, in particular, the SMEs, has underinvested in R&D relative to the food sector. This can be explained by a focus on commodities but this is an oversimplification insofar as any successful strategy in international markets requires adequate investment. The limited R&D commitment of Irish dairy companies was reinforced by a spokesperson for a research centre, who indicated that indigenous Irish companies are not research driven. We therefore conclude that R&D strategies can be identified as a weakness in the Irish dairy diamond.

Managerial motivation. A final consideration is the managerial capability and motivation which underpins strategies. An important determinant of individual behaviour and effort is the reward systems under which employees operate (Porter, 1990, p. 113). Thus interviewees were asked to outline the basis for senior management incentives (Table 3.22). The principal conclusion is that management in most companies are motivated and driven by incentives. While the form of these incentives may vary, they all share a short term perspective. The question for the future is whether a longer term perspective is required. In conclusion, the impact on competitiveness, of managerial motivation, has been substantial as management have successfully responded, to the short term pressures, by reducing costs as reflected in the increased profitability reported in Table 3.1.

3.6.2 Domestic rivalry and co-operation

Porter argues that vigorous domestic rivalry among firms based in the same national economy may create and sustain international competitive advantage

Table 3.22 Basis for incentives in the Irish dairy processing industry ($n = 11$)

Basis for incentives	Number of companies	
	Small	Large
No incentives	2	0
Related to milk price	1	1
Performance related to profit and goals	1	3
Non-financial – related to performance	2	0
Incentive but not specified	1	0

Source: Based on company interviews.

in an industry. In the Irish market, this competition was confined to supplies of milk with minimal significant competition in the market place. Consequently, innovative efforts were focused on cost reduction rather than market or product development.

Competition in the Irish market place is largely confined to liquid milk and some consumer products. It tends to be intense but quite regionalized. The domestic markets for commodities and ingredients are considered too small to have a significant impact on competitiveness. Foreign competition is not seen as a major force, except by the larger companies and mainly in the spreads market. However, although all companies stated that competition in the Irish market is very intensive and largely related to price and quality it was not considered a significant factor in the explanation of competitiveness. The only exception may have been in the development of the spreads sector where extensive 'copy-catting' may have led to Irish companies becoming heavily involved in spreads. Most companies were in agreement that rivalry in the foreign markets is more intense. Finally none of the companies claimed that the rivalry had given rise to major innovations on the product side.

The other aspect to rivalry is competition for supplies. Several studies (for example O'Dwyer, 1970; Keane and Pitts, 1981; Igoe, 1993; Gill, 1995) have shown that a consistent feature of the Irish dairy processing industry is that there are too many companies. However, Keane (personal interview, 1996) suggests that the intensive rivalry on the supply side may have been the impetus for increases in processing efficiencies. In the 1980s companies competed aggressively for supplies although this is now less prevalent. This form of rivalry represented a short term pressure for companies as they competed to pay a higher price for milk. The lack of impact on new product development and innovation was dramatic and is captured in the response 'who would thank you'. Further, there has been no real pressure in the last 10 years to alter that position as the Irish dairy processing industry had a guaranteed or supported outlet for most of its produce which facilitated continued payment of a high price for milk. The processors focused on the cost side in order to compete. The industry focused development on cost-effective technically efficient operations with extremely cost conscious management in a European context but behind New Zealand. Some companies agreed that this cost focus may have enabled the Irish companies successfully to pursue their international acquisition strategy in the UK and US. The rivalry also reduced take-over opportunities in Ireland, thereby forcing companies to look beyond Ireland. Companies believed that it was often not possible, and certainly more difficult, to achieve sufficient scale in Ireland.

Managers were also asked to comment on the level of co-operation with other dairy companies. We looked into three areas of co-operation: production, marketing and research and development. Apart from the earlier mentioned 'milk deals' and some, often short term, contract arrangements involving the production and/or packaging of milk products, we found no evidence of 'real' production arrangements (joint ventures, licensing or franchising) between Irish dairy processors. The latter kind of production co-operation only takes place with competitors abroad, and is almost exclusively limited to the larger Irish processors.

As regards marketing, most of the companies mentioned the co-operation through their stake in the IDB. Apart from this we found no examples of direct marketing co-operation between Irish dairy processors. Some of them have forged co-operative marketing relations, but always involving foreign dairy processors and almost exclusively for the marketing of Irish products in foreign markets rather than the marketing of foreign products in Ireland. We found some evidence of informal co-operation between Irish companies in export markets. In these markets companies try to prevent 'quoting for the same deal', thereby increasing their bargaining power.

Finally, regarding research and development, most managers mentioned their collaboration in Moorepark Technology Limited (MTL). Apart from this we found little evidence of direct research and development co-operation between Irish dairy processors. There is, however, informal co-operation, mostly linked to problem solving. Managers regularly consult each other on issues regarding production processes, equipment and engineering and will help with inventory shortages.

This kind of co-operation is often based on personal relationships, involving a high level of trust and integrity. The Irish dairy processing industry is a small industry where all the players know each other. Managers gave evidence of strong social and commercial interaction between competitors. Meetings at ICOS, the IDB and golf, among others, were seen as important for the creation of a valuable social network.

So, although there is a substantial amount of interaction between the Irish dairy processors, there does not appear to be much direct co-operation. There is no 'real' co-operation between Irish companies in the area of production. Co-operation is most developed in the area of marketing where the processors have been able to increase their marketing scale and bargaining power through the operation of the IDB. Most managers thought that this co-operation had been beneficial for the competitiveness of the Irish dairy industry. The informal marketing 'co-operation' between Irish companies in export markets was also seen as beneficial. Finally, in the area of research and development, managers acknowledged the role of indirect co-operation. However, the co-operation in technical problem solving, although helpful, is not likely to have a significant impact on innovation.

Thus, in the Irish dairy processing industry we found elements of both competition and co-operation. In most cases, the type of co-operation identified is of an indirect nature. The impact of competition is more obvious, although the few examples of indirect forms of co-operation have had a limited but positive impact. However, on balance there appears to be more competition than co-operation.

3.6.3 *The importance of strategy, structure and rivalry*

Has the level of rivalry and the strategy and structures which have been adopted facilitated the development of a competitive Irish dairy industry? Notwithstanding some reservations, it is reasonable to conclude that, within the regulatory framework, the structure and strategies of the Irish dairy industry have had a positive impact on the development of competitiveness. The industry has been dominated by a co-operative ethos and a production orientated philosophy. The emphasis has been on commodities and butter in particular. The IDB has enabled many companies to survive and the ensuing rivalry, in parallel with substantial interaction among the companies, has had a positive impact on competitiveness. The companies have successful and aggressive expansion strategies. They have broadened their product portfolios and differentiated their products by various means but are still oriented towards commodities with value arising primarily through relationships. The management have performed effectively and are responding to various short term pressures to maintain that performance. The international investments made have concentrated on gaining greater access to resources, markets and efficiencies. However, R&D strategies are arguably a weakness of the sector.

3.7 The CAP and the Irish diamond

So far we have not dealt with a final variable, influencing the four determinants of the Irish diamond, the EU CAP. *Prima facie*, the entire EU market is regulated by the same regime, and therefore impacts all competitors equally. However, we argue that the CAP, in conjunction with national diamonds, will have different effects in individual member states and may therefore be a significant variable in the explanation of competitiveness of the Irish dairy industry.

3.7.1 *The European CAP*

In Ireland, as in the rest of the EU, the dairy industry operates in a highly regulated environment. Since 1973, the CAP has been the most influential policy operating on the Irish industry. Through a system of intervention support, prices for dairy products in the EEC (as it then was) were kept well above world prices. The EEC market was protected from imports by import levies and exports were made competitive on the world market through export subsidies. This market support system provided a guaranteed outlet for dairy products at attractive prices. This regulatory regime was a major factor in the growth in the production of dairy products during the 1970s and into the early 1980s.

As production grew throughout the EC in the 1970s and 1980s, problems arose with surpluses of dairy products and the cost of the market support measures. As a result, quotas or limits on milk production were introduced in 1984. In addition, changes in the intervention system were introduced to make it less attractive as an outlet. Delays in the payment for produce placed in intervention as well as a tendering system, which effectively reduced the intervention price, were introduced. Continuing surplus production and budgetary problems along with pressure from the Uruguay round of GATT negotiations resulted in further reform of the CAP in 1992.[14] The net effect of the reform of the CAP has been to reduce and limit the level of the raw material available to the dairy industry. In addition, the sale of butter and skimmed milk powder into intervention has been made less attractive.

3.7.2   The CAP and the Irish dairy industry

The impact of the CAP in the Irish dairy industry was not the same as in other EU member states, partly because of the favourable treatment the industry received in CAP reforms. In 1984, with the introduction of the quota system, quotas were generally fixed at 1981 national deliveries plus 1%. However, Ireland's quota level was fixed at 1983 deliveries plus 4.6%. This favourable treatment continued in consecutive rounds of negotiations through the 1980s when national quotas were cut. Between 1985 and 1995 the Irish quota was cut by only 6.3%. This compares favourably with cutbacks in Denmark (9.7%), France (9.5%), Netherlands (9.2%), 'West' Germany (8.7%) and the UK (8.5%) (ABN.AMRO, 1995, p. 38).

The favourable outcome of successive rounds of quota negotiations was that Ireland's share of EU milk production increased.[15] This, combined with a guaranteed market and export subsidies, led to a higher share of total EU exports. It has been argued that the outcome also affected other indicators of performance, such as overseas acquisitions and value added. For example, the reduction in the UK quota has turned it from being nearly self-sufficient in milk into a country in net deficit by about 10%. In contrast, Ireland's quota still leaves it with a substantial surplus relative to local consumption. Others have argued that this may have been one of the reasons why Irish dairy companies were able and willing to acquire some UK dairy operations, for further processing of Irish dairy produce (ABN.AMRO, 1995).

Thus, the 'competitiveness' of the Irish dairy industry, as expressed in its share of world exports, has been partly facilitated by a regulatory regime that has discriminated in favour of the Irish dairy industry. It is tempting to conclude that the national diamond is of little explanatory value for the competitiveness of the dairy processing industry. However, we argue that the CAP has worked in conjunction with the Irish diamond to create a competi-

14. Under this reform package, the support price for butter is reduced by 5% from 1993 to 1995 and milk quotas are reduced by 2% in the same period.
15. Note that the Irish share in EU milk production actually dropped slightly, due to the reunification of Germany and the consequent increase of the total EU-12 quota figure.

tive industry. Below we start by showing that the Irish dairy industry has been able to influence CAP policies and, furthermore, show that, given the characteristics of the Irish diamond, the specifics of the CAP suited the Irish dairy industry.

Kamann and Strijker (1995) rightly point out that the EU CAP is not something autonomous, something given:

> The outcome and nature of the regulatory regime in any market is politically determined and therefore is subject to influence from lobbies by interested parties. (p. 107)

It is our view that the 'favourable treatment' of Ireland in consecutive CAP negotiations was not given, but won. The Irish have proved to be very effective in negotiating concessions for their dairy industry:

> The small 6.3% cutback in Ireland is perhaps a reflection of the effectiveness of Irish political lobbying in a country where dairy farming is proportionately more important to the Irish economy (ABN.AMRO, 1995, p. 38)

This effectiveness, in turn, could be seen as reflecting the continuing importance of agrarian interests in Irish politics relative to other European countries.[16]

Furthermore, the CAP suited and reinforced particular characteristics of the Irish dairy industry (i.e. production of commodity type products for export markets and a relatively fragmented industry structure) and allowed them to achieve their corporate objective of survival. Commodity-type products have always constituted a large part of the product portfolio of the Irish dairy industry. The combination of seasonality, a small domestic market, peripheral location and being an industrial latecomer, resulted in a commodity oriented production.

Climate and other natural conditions made the seasonal production of milk (and dairy products) economically appealing. Relative to milk production, the domestic market was small, and this created a surplus for exports. Export of commodities was the most feasible option. The requirement of a year-round milk supply, for the manufacture of branded consumer products, together with the country's peripheral location, made the production of storable commodities for export a logical option. Furthermore, being an industrial late-comer, the Irish dairy industry lacked international brands which would be difficult to develop by relatively small companies in a fragmented industry.

The specifics of the CAP support system suited this situation. The system provided a guaranteed support for storable commodities (butter and skimmed milk powder) which were allowed to be produced in a seasonal pattern. The CAP also reinforced some characteristics of the Irish dairy industry. Seasonality was reinforced because of the existence of a guaranteed market,

16. Mjoset (1992) presents a comprehensive analysis of the reasons for the continuing importance of agricultural interests in Ireland relative to other European countries.

which took away the 'risk' of seasonal oversupply. Furthermore, the product portfolio became more orientated towards commodity-type products because the home market for (branded) consumer products was nearly saturated and most of the scope for increasing output was in the area of commodities for exports.

The CAP worked against long term development of export markets for unsupported products. Farmers are a risk averse group and their strategic horizon is shorter than what would be preferable for processors (Harte, 1995). As described in the section on rivalry, the Irish processors, have been under pressure from farmers to pay the best price for milk, a rivalry stronger than in other countries. The CAP reduced the need for 'risky' investments in branded products because it offered a guaranteed market for commodities. This, in combination with the rivalry and the relatively short-term strategic outlook of the farmers, meant that management of the farmer-controlled processors could count on less support for high risk investments in products and markets.

Finally, the CAP also influenced the structural development of the industry. In this research a number of interviewees argued that the CAP system reduced the need for consolidation in the Irish dairy processing industry which would have been necessary in an unregulated market. Because of the possibility of selling commodities into intervention there is less need to develop new products for unprotected markets, which would require larger companies. This allowed the relatively fragmented production structure in Ireland (as well as in some other EU member states) to survive.

The structure of the industry was further affected by the introduction of quotas. The quotas were an additional barrier to entry. Quotas have limited the milk available for processing and this allowed some processors to sell part of their milk to the highest bidder, while relying on others to invest in processing and markets. This further reduced the pressure for rationalization in the industry.

3.7.3 Sustainability and the CAP

Thus, the Irish diamond worked in conjunction with the CAP, to create a competitive industry. However, while the industry might be considered competitive in the present environment, the sustainability of this competitiveness is questionable. The CAP has left the industry dependent on (directly or indirectly) supported, storable commodity products. While one could argue that the commodity orientation of the industry is bound up with seasonality, peripherality, and a small domestic market, the fact remains that it was the CAP that allowed for a relatively risk-free increase in production of supported commodity products, thereby increasing the commodity orientation.

With the prospect of continued deregulation and declining support for the dairy sector under the CAP regime, the industry is left with a product portfolio which is likely to come under increased competitive pressure. Furthermore, it is questionable whether the surviving structure of the dairy processing industry is sustainable in an unregulated and unsupported market.

Declining support for EU export products, such as butter and skimmed milk powder, will mean that Irish producers will face strong competition from efficient commodity producers in, for example, New Zealand and the Netherlands. Although it is not impossible for Irish dairy processors to become very competitive commodity producers, able to face such competition, such a route would leave the Irish producers vulnerable to world market fluctuations. Furthermore, in the long run, such a route is unlikely to 'increase the standard of living for the citizens of Ireland' (Porter, 1990, p. 6). Shifting all these products towards the European market will be very difficult because of the mature nature of that market (Kamann and Strijker, 1995). The option of utilizing the available milk for the production of less mature and/or branded products will be difficult given the lack of developed brands. And, for reasons explained earlier, it is unlikely that the co-operatives will make substantial investments in the development of such brands.

This is not to criticize the co-operative ethos. On the contrary, the co-operatives are successfully performing their primary task: to sell all the milk of the owner-farmers, in whatever processed form, for the highest price, with low risk. Although the farmers, controlling the processing side of the industry, are very aware of the future changes in the CAP and the possible requirements for sustainable competitiveness of the processing side of the industry, they seem reluctant to support major investments in product and brand development.

This is not an irrational move, but a strategic decision based on their entrepreneurial interest in the farm and its revenue. For the farm, as a business, it simply does not make sense to sacrifice the short term gains of a higher milk price for the (high risk) development of future products and markets that might not be there, precisely because of the changes that make these investments necessary in the first place. The Irish dairy processing industry is trapped in a CAP web. The future changes in the CAP regime require actions that are not likely to be taken before the changes are actually taking place. A similar view is held by Kamann and Strijker (1995) in relation to the Dutch dairy processing industry.

3.7.4   Influence of the CAP on the Irish dairy processing industry

It is clear that the CAP has positively influenced the performance of the Irish dairy processing industry. The competitiveness of the Irish dairy industry has been facilitated by a regime that has discriminated in its favour. Furthermore, the specifics of the CAP support system suited and reinforced the particular characteristics of the Irish dairy industry and allowed them to survive. However, the role of the CAP has been more than just 'an outside influence on the four determinants' of the Irish diamond (Porter, 1990, p.

127).[17] The EU CAP is not just something 'given', influencing the diamond from outside. The Irish dairy industry has been able to influence the CAP policies and its outcomes. It is our view that the CAP has not just influenced the Irish dairy industry, but worked in conjunction with the Irish diamond to create a competitive industry.

Thus, the Irish industry has used and influenced the CAP regime to its advantage to become an important exporter of dairy products. However, while the industry might be considered competitive in the present environment, the sustainability is debatable. With the prospect of continued deregulation and declining support for the dairy sector under the CAP regime, the industry is left with a product portfolio which is likely to come under increased competitive pressure. Although, it is not unthinkable that the Irish dairy industry could become a very competitive producer of basic commodities, a strategy focusing on the efficient production of such products is unlikely to create increased wealth for the Irish citizens in the long term. The latter objective requires substantial investments in products and markets, but the actions necessary for the sustainable development of the Irish dairy processing industry, trapped in its CAP web, are unlikely to be taken before the changes in the CAP regime actually occur. This is explained by the different time horizons of farmers and processors, together with the fact that it is unlikely that the same number of farmers would enjoy the fruits of present sacrifice.

This is not a critique of the co-operatives. The farmers, through their co-operatives, were (and still are) taking rational decisions. Comparing the performance of co-operatives with the (arguably more sustainable) performance of private dairy processors, does not make sense. The co-operatives have to take account of the farmers' interests. The only concern of private processors is to create a good return on invested capital.

3.8   Assessment and conclusions

To conclude, we first summarize the contribution of Porter's four determinants of competitive advantage to the performance of the Irish dairy processing industry. This is followed by an assessment of the significance of clustering in the creation of competitive advantage in the dairy processing industry. This will enable us to reflect on the suitability of Porter's theoretical framework for analysing competitive advantage in the dairy processing industry. Finally, we outline the policy implications of our study.

3.8.1   The role of the determinants of competitive advantage

Ireland has a reasonable endowment of factor conditions. The level of technical expertise, operational skills, the supply of graduates and a healthy

17. This statement refers to Porter's view on the role of government. We have treated the EU CAP as part of the regulatory regime (government).

national image were all identified as relatively strong elements. As regards the cost base, costs of logistics and, to a lesser extent, labour and energy costs have a somewhat negative impact on effective progress in the Irish dairy industry. Finally, marketing skills were identified as a negative factor. As regards the factor creating mechanisms, the Irish educational institutions and the supporting agencies or institutions were seen to play an important, positive, role. Overall, factor conditions in Ireland have an important, and positive, impact on the competitiveness of the dairy processing industry.

Home demand conditions have been less important for the competitiveness of the Irish dairy industry. Irish consumption patterns might have worked against the development of some segments of consumer products, such as yoghurt and desserts. However, while Ireland's main EU competitors all benefited from a broader range of often more sophisticated demands, the small absolute size of the home market seems to have played a bigger role than the concept of segmentation.

The retailers and foreign multinational enterprises located in Ireland have a positive influence on the competitiveness of the Irish dairy industry. The Irish multiples can be regarded as sophisticated and demanding buyers and they are, to some extent, responsible for upgrading the standards of the larger Irish dairy processors. However, a more important influence of the retailers has been their constant pressure on the processors to reduce prices, and the consequent effects on cost efficiency. The multinational customers located in Ireland have also played a positive role. Most processors learned from the standards and systems employed by these multinationals, an experience that helped them in international markets. It is, however, important to recognize that many processors are also fostering intensive and beneficial relations with multiples and multinationals located abroad.

Apart from the Irish dairy farmers, related and supporting industries in Ireland contribute little to the competitive advantage of the Irish dairy processing industry. The Irish dairy farmer is both a supplier and a shareholder of the Irish dairy processing industry. Because of this, it is difficult to determine the exact contribution of the farmers to the competitiveness of the processing industry. The data suggest that Irish farmers, compared to farmers in most other European countries, are competitive producers, but this advantage is partly offset by a highly seasonal supply.

There is very little evidence that the Irish processors benefit from the local presence of other (non-milk) supplier industries. Most of the strategically important machinery is imported. Although, there are some links with related industries, again the contribution to competitiveness is limited. Some suppliers appear to have benefited from indirect links with other processing industries which suggests potential for further development. Most of the interviewed managers did not perceive the lack of related and supporting industries as a competitive disadvantage. However, one could argue that it is difficult for managers to appreciate the potential benefits of absent related and supporting industries. Porter argues that the lack of competitive suppliers diminishes the potential for ongoing co-ordination, important for innovation and upgrading (Porter, 1990). This would place the Irish processors in a disadvantaged position.

As regards the contribution of firm strategy, structure and rivalry, one characteristic is the important role played by the co-operatives. Co-operation through the IDB partly overcame the disadvantage of the relatively small size of the Irish companies in international markets. The Irish dairy processors showed a willingness to adopt innovative governance structures, e.g. the co-op/plc structure. An examination of the strategies suggests that, often, a cost based competitive focus has been pursued. The companies have become active in international acquisitions and are actively attempting to add value. The primary mechanism is through the formation and maintenance of relationships, either directly or via the IDB. The companies are directing efforts towards growth in areas of competitive strength rather than in downstream activities.

Porter (1990) believes that the existence of rigorous domestic rivalry facilitates the creation and persistence of competitive advantage in an industry. Certainly, Irish companies have competed vigorously, mainly within concentrated geographic regions. However, this competition focused mainly on the supply side. We suggest that this competition for raw materials had a positive impact on competitiveness as processors had to focus on cost reduction. In parallel, co-operation exists at various levels. The companies collaborate and support the IDB, which is now a major international player in the European dairy industry. In a similar vein, MTL is evidence of joint research. Companies also collaborate informally on international marketing strategies. In most cases, however, the type of co-operation identified is of an indirect nature. The few examples of indirect forms of co-operation have had a limited but positive impact

The industry has organized itself into a structure consisting of processing companies, a group marketing organization (IDB) and a number of support organizations, e.g. ICOS, which have effectively represented the industry at government and supragovernmental level. In parallel the industry has competed internally to produce a much leaner industry capable of improving its international position. In conclusion, notwithstanding some reservations, it is reasonable to conclude that, within the regulatory framework, the structure, strategies and rivalry of the Irish dairy industry have a positive impact on competitiveness.

Finally, the CAP is posited, in the same way as 'government', as an additional variable, an outside influence on the four determinants. The CAP has positively influenced the performance of the Irish dairy processing industry. The regime discriminated in favour of the Irish dairy industry and the opccifics of the CAP support system suited and reinforced its particular characteristics. However, the EU CAP is not just a 'given', influencing the diamond from outside: the industry has organized itself to influence the CAP policies and its outcomes.

With the prospect of continued deregulation and declining support for the dairy sector under the CAP regime, the industry is left with a product portfolio which is likely to come under increased competitive pressure. Given the pending deregulation of the dairy markets, a strategy focusing on the efficient production of basic commodities is unlikely to create increased wealth for Irish citizens in the long term. Changing this focus is one of the most important challenges of the Irish dairy industry, but the necessary actions are not likely to be taken before the CAP changes actually take place.

3.8.2   Assessment of 'clustering'

Having identified the relative significance of the four determinants of competitive advantage we now need to identify and explain the significance, or non-significance, of clustering in the creation of competitive advantage in the dairy processing industry. We therefore need to consider whether there is evidence of a clustering process, and its significance for competitive advantage.

Figure 3.5 brings together the earlier identified horizontal and vertical linkages of the Irish dairy processing industry, which are characteristic of a

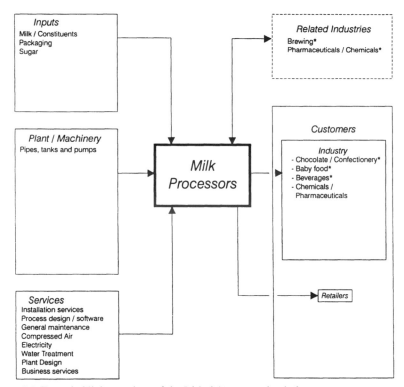

Figure 3.5 Extended linkages chart of the Irish dairy processing industry: outcome.

'cluster' in the sense used by Porter. The core of the Irish cluster is made up of a relatively large amount of competing indigenous milk processors, some of which have developed into competitive multinational companies with headquarters in Ireland. We found many vertical linkages and other commercial interactions between the competitors. In the area of marketing, the interaction between the processors has evolved into a form of co-operation.

Certainly the smaller co-operatives could be described as a 'network of processors supplying one sales organization'. However, all processors are free to sell their produce through other marketing channels, and on balance, there appears to be more competition than co-operation. We also found evidence of social interaction. In this regard, meetings at organizations such as ICOS and the IDB, among others, were seen as important for the creation of a valuable social network.

But is the dairy processing industry part of a larger cluster of competitive industries? We found some forward linkages with competitive customers, most notably with the baby food and confectionery industries. The presence of these industrial customers appears to have contributed to the competitiveness of the dairy industry. Almost all these industrial customers are multinational companies, with their home base outside Ireland. We also found that the relationships between processor and multinational do not always involve substantial interaction. Apart from that, some of these downstream industries involve only a limited number of companies. The second group of customers, the Irish retailers, have not developed into global players. In spite of this they had a positive effect on the standards and cost efficiency of milk processors.

On the input side, the processors are 'supported' by dairy farmers who are both suppliers to, and (in many cases) shareholders of, the processing co-operatives or companies. The dairy farmers form an important part of the cluster. For obvious reasons, the competitiveness of the Irish dairy farmers, compared to farmers in other countries, cannot be expressed in terms of world export share. Also, an assessment on the basis of milk prices proved inconclusive, precisely because of the strong link between supplier and processors. However, the data suggest that the Irish dairy processors are supported by competitive milk producers. Furthermore, the 'common ownership' structure has resulted in a strong supplier–customer interaction.

Apart from dairy farming, the number of competitive suppliers located in Ireland is extremely limited, and only the packaging industry could be regarded as a competitive industry in the Porter sense. Apart from some 'captive' engineering companies, supplying some basic equipment, there are no manufacturers of dairy machinery in Ireland.

Finally, two industries, brewing and pharmaceuticals have the potential of developing into related industries to dairy processing. The pharmaceuticals/chemicals, brewing and dairy processing industries use similar technologies and we found some examples of reciprocal direct interaction and information flow between these industries. We also found examples of indirect technology transfer via the suppliers, and the movement of technical staff between these industries. However, although these links might have benefited the three industries to some extent, most links seem to be incidental rather than part of a constructive network of related industries and until now the impact on competitiveness has been limited. Apart from that, both the brewing and chemicals/ pharmaceuticals industries are mainly foreign-owned.

We already saw how the different elements of the Irish diamond have influenced the competitive advantage of the dairy industry. But has the Irish dairy 'cluster' promoted the interaction between the four elements? In other words, is the Irish diamond operating as a system? There is some support for this. First the presence of a number of strong competitors in the dairy processing industry has instigated the establishment of two high-standard factor creating research facilities developing generalized and advanced technology, MTL and the research facilities at University College, Cork. Furthermore the strong demand of the dairy processing industry together with the demand of other processing industries for skilled labour resulted in a wider range of courses in educational institutes and increased the availability and standards of skilled labour.

The cluster of milk processors in Ireland has also affected demand conditions in the Irish market. The availability of (primary processed) milk formed an important attraction for a number of multinational companies to set up production facilities in Ireland. In turn, these companies increased the demand for skilled labour, the sophistication of production facilities and the standards of the Irish processing industry in general. Finally, there is some evidence that the presence of the milk processing sector and other related processing industries led to the formation of new firms and skills in the supply sectors, although the number of spin-offs from the processing industries has been extremely limited.

Although these points can be interpreted as evidence for a (potential) 'diamond as a system', the question remains whether the functioning of the 'diamond as a system' and the process of clustering have been facilitated by geographic proximity. The location of the biggest Irish dairy processors and some of the most important supporting industries and organizations illustrates that Ireland's dairy cluster appears to have a strong geographical dimension. The economic activities of dairy farming, dairy processing, and the few supplier industries are all spatially clustered in the heartland of Irish dairy farming. Furthermore, while many of the institutions identified as important (Department of Agriculture, Food and Forestry, ICOS, IBEC) are located in the Dublin area, the most important dairy related research institutes (University College, Cork and Moorepark Research Centre) are located in the south.

The reasons for this spatial concentration are apparent. Because of favourable climatic conditions, the heartland of Irish dairy farming developed in the southern province of Munster. The perishable and low value/weight ratio of the milk supply acted (and still acts) as an important locational determinant for the processing industry with the result that the processors showed the same regional concentration. Transport costs were again the main locational determinant for some multinationals using the output of the processors, which also located in the same region, and in some cases on the same premises as their suppliers.

The geographical proximity of the engineering companies seems more likely to be the outcome of a selection process of the processors rather than of the location decisions of the engineering companies themselves. Finally, the location of the main research institutions is the outcome of planning decisions which were influenced by the location of the processing industry.

It is clear that the spatial concentration of competing processors is a result of locational constraints in relation to raw material supply, rather than reflecting the need for interaction (or information flow) between the processors. However, there is evidence of an agglomeration effect where a cluster of processors to attracted suppliers, customers and supporting organizations to the same region.

Porter identifies domestic rivalry as a second element with especially great power to transform the diamond into a system, because it promotes upgrading of the entire national diamond. The Irish processing industry is characterized by substantial rivalry, particularly in the context of raw material sourcing. We failed to find much evidence to confirm Porter's proposition that this would lead to the upgrading of the other elements of the diamond and have a consequent impact on innovation. The Irish experience in the area of dairy spreads, characterized by several companies trying to gain market share by introducing new product variations, was the only obvious example of a possible systematic impact of rivalry.

On the other hand, we found examples of co-operation between processors that could be argued to have had an impact on the other elements of the diamond. In the area of export, the joint marketing effort in the form of the IDB facilitated the development of innovations such as an international brand name. This would have been less likely in a situation of non-co-operation. Furthermore, collaboration through MTL has upgraded the level of knowledge available to the dairy industry. Finally, the collaboration between the different co-operatives also led to the formation of a co-operative supplying packaging to the dairy industry.

From the above it is clear that, on a number of points, the Irish dairy processing industry falls short of being a cluster in the Porter sense of the term. First, the cluster is limited in scale and scope. Although Ireland has a relatively large number of dairy companies, the number of non-milk suppliers, customers and supporting institutions is very small and the scope of their activities is limited. Second, many Irish suppliers and customers are not part of a competitive indigenous industry themselves. Third, most of the important downstream and related industries involve mainly foreign-owned multinational companies with their home base outside Ireland. Fourth, although we found some evidence of beneficial interaction between milk processors and customers increasing the competitiveness of the milk processing industry, the benefits do not seem to be reciprocal. In other words, the interactions are not always part of 'a mutually reinforcing process'. Finally, although we found evidence of geographical concentration, a large part of the business within the cluster takes place in the absence of substantial direct interaction. So, although the Irish dairy processing industry has some characteristics that might be indicative of a 'clustering process' in the Porter sense of the term, an Irish dairy cluster supported by an Irish diamond functioning as a system leading to innovation and sustained growth, has yet to develop.

3.8.3 Suitability of Porter's theoretical framework

Porter's diamond has proved a useful framework for analysing the dairy industry. It enabled us to identify the factors that explain the competitiveness of the Irish dairy industry. However, the findings fail to support all of Porter's propositions to the same extent. The relative success of the Irish dairy industry has been achieved in spite of the underdeveloped nature of some elements of the diamond and in spite of limited system dynamics. Furthermore, the importance of some influences seems to be underestimated by Porter.

First, domestic demand conditions have played a smaller role in determining the performance of the Irish dairy processors than Porter's model would suggest. The presence of the few 'sophisticated and demanding buyers' has been helpful but far from crucial. Furthermore, contrary to Porter's ideas, the small size of the Irish home market has proved a competitive disadvantage for the production of certain consumer products. Second, apart from the modestly positive impact of the dairy farmers, the processing industry has not benefited from the presence of competitive local suppliers. Third, although the 'competition' for raw milk supplies had a positive impact on efficiency. the impact of competition for (regional) market share appears to be limited. Finally, although the Irish dairy processing industry has some characteristics that might be indicative of a clustering process in the Porter sense of the term, the industry has been able to compete in the absence of a fully developed 'dairy cluster' and its associated system dynamics.

Apart from this, Porter fails to fully acknowledge a number of issues that proved important in explaining the competitive advantage of the Irish dairy processing industry. First, although Porter lays stress upon strong rivalry between individual firms, different forms of limited, indirect, co-operation between competing processors seems to benefit the Irish dairy processing industry. Second, subsidiaries of foreign multinationals in Ireland have played a more positive role than would be expected on the basis of Porter's model, both as key customers and as related companies. Finally, by treating the CAP as an additional variable – 'influencing the four determinants from outside' – we have succeeded better than Porter in capturing first, the influence the Irish dairy industry has on the policy making process and, second, the difference in ability of national industries to react to specific policy outcomes.

Most of the above points have been identified before by critics of Porter's diamond model. Several critics consequently pointed to Porter's own qualification to his model – that one can succeed in resource based industries with only one of the elements of the diamond in place (e.g. Yetton *et al.*, 1992). Some critics argue that the model is not applicable to resource based industries (such as the dairy processing industry) and again claim to find support in Porter's writing:

> The capacity to compete in industries that are highly dependent on natural resources might be more explicable using classical theory. (Porter, 1990, p. 28)

However, Porter does not intend to limit the applicability of his model to non-resource-based industries. Porter acknowledges the fact that firms in a country may sometimes be able to sustain their competitive advantage for a time, based solely on the country's natural resources. However, by this, he does not mean to invalidate his model for resource based industries. For example, in responding to critics, Porter and Armstrong (1992) explain that there is nothing wrong with resource industries being the mainstay of an economy. However, they criticize countries that export only unprocessed or semi-processed products with little presence in more sophisticated segments. The challenge that they identify is to increase the sophistication of the way in which countries compete in natural resources industries through more efficient production and through migration into more sophisticated segments. It is in relation to this challenge that Porter defends the relevance of the diamond model, for (natural) resource based industries.

It is our view that Porter's diamond model and the notion of clustering are relevant concepts in relation to the Irish dairy processing industry. Although the Irish dairy processing industry has prospered in the absence of a fully developed 'dairy cluster', there are a few reasons for supporting the concept of clustering and its importance for competitive advantage in the dairy industry. First, although an Irish dairy cluster has yet to develop, we found some evidence that was indicative of a clustering process and some elements of the diamond have developed and are playing a positive role. Certainly, the Irish dairy industry is not relying on basic factors only. Raw milk does not constitute a factor-based comparative advantage that, following classical rationale, could be enough to lead to a strong export position in commodity dairy products. The section on dairy farming showed that, taking account of the composition of milk and the seasonality of supply, the Irish dairy processors are not deriving an extremely strong advantage from the local milk supply

Second, as mentioned earlier, it might be difficult for individual players in a cluster to perceive all the benefits of a clustering process and the benefits might, therefore, be greater than perceived by the interviewees. Furthermore, it is difficult, in the absence of a fully formed cluster, to recognize the potential benefits of a fully functioning diamond.

Finally, although the performance measures presented in this chapter suggest that the Irish dairy industry is performing well, Porter might have some reservations in calling it a competitive industry. The European, and indeed many of the international, dairy markets are extremely protected and regulated. Part of Ireland's positive performance is the result of sales in protected markets, sales into intervention and subsidized exports. The Irish dairy industry is not really 'competing' with competitors from, for example, New Zealand.

Furthermore, we saw that the specifics of the CAP support system, i.e. support for basic commodity type products, suited the Irish milk processors. However, we also saw that the pending changes in the CAP and GATT might provide a new challenge for Irish processors. The challenge is to increase the sophistication of the way in which they compete in a natural resource-based industry, through more efficient production and, more important for the wealth of Ireland's citizens, through migration into more sophisticated segments (which could mean more sophisticated commodities). In this situation it will be beneficial to have at least a number of the elements of the diamond in place, acknowledging that it may be possible to prosper without having all the elements developed to the same extent.

3.8.4 Implications for policy

The Irish dairy processing industry has been able to prosper without the presence of a fully developed diamond. However, as discussed above, pending changes in the CAP and GATT regimes will inevitably increase the competitive pressure on Irish processors, particularly in basic commodity markets. If the Irish industry is to compete effectively in the new environment, then steps need to be taken to further develop the Irish diamond which will allow the industry to develop a competitive advantage in a less regulated and supporting environment. However, in prescribing policy recommendations we must take into consideration some of the qualifications identified in the preceding sections.

The capacity to innovate will to a large degree depend on the availability and quality of labour. The level of operational skills and the supply of graduates are identified as relatively strong elements of the Irish diamond. It is important that this position is maintained and further developed. According to Porter and Armstrong (1992), education and training constitute perhaps the single greatest long-term leverage point to all levels of government in upgrading industry. But, in certain areas, firms must accept and play their own role in education and training.

As regards knowledge and expertise, it is important that processors invest in expanding the existing base and increase the efforts in product development, particularly in technology-driven market segments, such as ingredients and functional food products. Although, we see an important role for the Irish government in stimulating institution-based R&D, again, processors must play their part. Government should continue to finance competitive and basic research; however, it is important that the benefits of these efforts are accessible to the industry as a whole rather than to individual companies. In this respect there will be a greater need for closer integration of research institutions and industry. It is our view that indirect co-operative research can be beneficial provided it takes place in research institutions to which the majority of industry participants have access. The government should continue in its efforts to facilitate this type of co-operation.

Furthermore, we advocate the government's involvement in stimulating a deepening and broadening of the existing dairy cluster. However, in this process one should be realistic and appreciate the constraints of such a strategy. For example, the scope for development of backward linkages might be limited. The machinery sector is dominated by a small number of large international firms. In such a concentrated industry the barriers to entry might well be too strong for new indigenous companies to develop successfully. In some cases, one might conclude that for a small economy it is more practical to attempt to capitalize on expertise of suppliers abroad. Thus, policy should not attempt to 'fill all the gaps in the cluster'. However we identified a potential for the further development of service companies that might benefit a broader 'processing cluster' in Ireland.

There are, however, more obvious possibilities for a further development of downstream linkages. This potential could be developed through related diversification by the existing processing companies in Ireland and, also, through the targeted attraction of foreign food companies. According to Porter and Armstrong (1992), a country should seek to attract foreign multinationals that operate in industries within those broad sectors in which the nation's firms might themselves eventually gain competitive advantage. Here, multinationals can seed a cluster. However the focus should be on the development of indigenous companies. Furthermore, policy should aim to facilitate more effective information flow between dairy companies, customer industries and related industries. On an international scale, the IDB must continue to give priority status to linking processors with customers abroad.

The future success of the industry is dependent on the ability of the management of the Irish dairy industry to adapt their structures and strategies to suit the changing environment. Managerial behaviour is affected by the process of corporate governance. According to Porter, governance structures in which the boards represent the interests of investors, and in which investors have a role in management, would normally lead to more emphasis on building long term shareholder value. However, in the case of the Irish dairy processing industry the situation is complicated by the fact that the shareholders are also the suppliers of the most important raw material. We identified the resulting 'horizon problem' as an impediment for the long-term development of the Irish dairy processing industry.

Furthermore, we found widespread acceptance for the view that there is scope for consolidation. This will, at least in the short term, result in job losses in the industry. Many consider the social and employment consequences too severe to justify change in the present climate. These are highly emotive and political issues which may require a stronger role of government.

Many of the above points have been identified by the different bodies responsible for policy making in Ireland. Indeed, the needs for rationalization, scale and diversification in the industry and for a shift to value added products have been consistent themes in the large number of policy documents relating to the dairy industry. Also, the need to reduce seasonality has been continually identified, as an impediment to diversifying the product range of the dairy industry. Forbairt has embraced Porter's cluster concept in its latest Food Development Strategy (Forbairt, 1995). The role of government in creating national advantage is significant, but inevitably partial. Without the co-operation of the industry, the best policy intentions will fail. All the major players in the dairy cluster, are aware of the need for action. However, the combination of the current governance structure and the pending changes in the CAP and GATT make it difficult for the industry to take the necessary steps. Difficult, tough it may be, it is important that action is taken before the changes in the CAP and GATT are actually in motion.

Acknowledgements

The research for this case study has been part funded by grant aid under the Food sub-programme of the Operational Programme for Industrial Development which is administered by the Irish Department of Agriculture, Food and Forestry and supported by national and EU funds. The study also formed part of a larger study 'Clusters in Ireland : a study of the Application of Porter's Model of National Competitive Advantage to Three Irish sectors', prepared by the authors and colleagues in University College Dublin and in The Economic and Social Research Institute for the National Economic and Social Council (NESC).

References

ABN.AMRO (1996) *Raising The Value-Added: A Strategic Review of the Top 15 European Dairy Manufacturers, 1996.* Dublin.

Clarke, A. (1995) *Operational Programme for Industrial Development; Final Report on the Food Sub-Programme 1994–1999.* Industry Evaluation Unit, Dublin.

Collins, A. (1996) *Evolving Channels: Changing Retailer-Manufacturer Relationships.* Discussion Paper No. 24, Structural change in the European Food Industry, University of Reading.

Cuddy, M. and Keane, M.J.(1990) *Ireland: a peripheral region, in The Single European Market and the Irish Economy* (eds J. Foley and M. Mulreany), Institute of Public Administration, Dublin.

Cunningham, S. and Pitts, E.(1995) *The Markets for Dairy Ingredients in the European Union*. National Food Centre, Dublin.

DMBI (1996) *European Logistics Survey* (Survey performed by Logistics Consulting Group), Danish Ministry of Business and Industry, Copenhagen.

Dunning, J.H. (1992) The competitive advantage of countries and the activities of transnational corporations. *Transnational Corporations*, 1, 135–68.

Foley, J. (1993) The Irish dairy industry: a historical perspective. *Journal of the Society of Dairy Technology*, 46(4), 124–38.

Forbairt (1995) *Food Development Strategy 1995–1999: An Action Plan for Growth*. Forbairt, Dublin.

Forfás (1995) *Research and Development in the Business Sector: Findings from the 1993 Census of R&D Performing Enterprises in Ireland*. Forfás, Dublin.

Forfás (unpublished extracts) Census of R&D Performing enterprises in Ireland. Forfás, Dublin.

Forfás (unpublished data) Irish Economy Expenditure Survey. Forfás, Dublin.

Gill, J. (1995) *The Irish Food Sector – A Case for Consolidation*. Riada, Dublin.

Goodbody Stockbrokers (1995) *Food Retailing in the Republic of Ireland*. Goodbody Stockbrokers, Dublin.

Harte, L.N. (1995) Creeping privatisation of irish co-operatives: a transaction cost explanation. Paper presented at the EIASM Workshop on Industrial Changes in the Globalised Food Sector, Brussels, 27–28 April.

IBEC (1995) *The Irish Food Sector: Competitiveness with the UK*. Irish Business and Employers Confederation, Dublin.

IDB (1995) *Annual Report 1995*. Irish Dairy Board Dublin.

Igoe, L. (1993) *The Irish Dairy Sector: An International Perspective*. Goodbody Stockbrokers, Dublin.

Jacobs, D., Boekholt, P. and Zegveld, W.(1990) *De Economische Kracht van Nederland* (The Competitive Advantage of the Netherlands) SMO (in co-operation with TNO), The Hague.

Kamann, D. and Strijker, D. (1995) The Dutch dairy sector in a European perspective, in *The Dutch diamond: The Usefulness of Porter in Analysing Small Countries* (eds P.R. Beije and H.O. Nuys), Garant, Leuven-Apendoorn.

Keane, M.J. (1995) An analysis of key factors affecting performance of Irish dairying. Unpublished PhD thesis, Trinity College, Dublin.

Keane, M.J. and E. Pitts (1981) *A Comparison of Producer Milk Prices in EEC Countries*. An Foras Taluntais, Dublin.

Mjoset, L. (1992) *The Irish Economy in a Comparative Institutional Perspective*. Report No. 93, NESC, Dublin.

MMB (1995) *EC Dairy Facts and Figures*. Milk Marketing Board, Surrey.

Nixon, C. (1995) *Value Added in New Zealand Agriculture: A Survey*. NZ Institute of Economic Research, Wellington, New Zealand.

O'Dwyer, T. (1970) The structure and organisation of a food industry, in *Into Europe – The Challenge to the Food Industry*. Conference Proceedings. An Foras Taluntais, Dublin.

O'Sullivan, A. (1982) *A development strategy for the Irish food industry*. Seminar Proceedings, NBST/IFSTI, Dublin.

PA Consulting Group (1996) *Developing Ireland's opportunity in dairy ingredients: sector study on future customer expectations*. Report to The Irish Food Board, Dublin.

Perez, R., Tozanli, S. and Vieille, J.N.(1994) *L'Industrie du Lait et ses Dérives en Europe* (Milk and Milk-Derivatives Industry in Europe). Eurostaf, Montpellier.

Pitts, E. (forthcoming) *A Strategy for Fats*. National Food Centre, Dublin.

Porter, M.E. (1990) *The Competitive Advantage of Nations*. Macmillan, London.

Porter, M.E. and Armstrong, J.(1992) Canada at the cross-roads: dialogue. *Business Quarterly*, 56, 6–10.

Varley, M. (1991) The Irish dairy industry: strategic options. Unpublished MSc Thesis, University College, Dublin.

Yetton, P., Craig, J., Davis J. and Hilmer, F.(1992) Are diamonds a country's best friend? A critique of Porter's theory of national competition as applied to Canada, New Zealand and Australia. *Australian Journal of Management*, 17(1), 89–119.

ZMP (1995) *ZMP Bilanz Milch 1995*. Zentrale Markt- und Preisberichtstelle GmbH, Bonn.

4 Uncompetitiveness in a primary product: does Porter help? The case of UK horticulture

BRUCE TRAILL

4.1 Horticulture in the UK

By any count, the UK horticultural[1] industries are uncompetitive, with imports easily exceeding exports (see Figures 4.1 and 4.2).

This is not surprising, given the country's climate and relatively high population density – a classic case of comparative disadvantage. One might

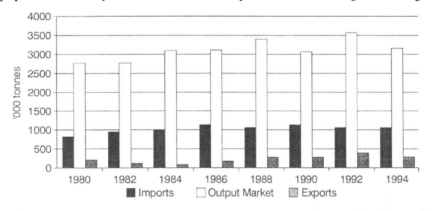

Figure 4.1 Comparison of the quantity of fresh and dried vegetables imported into the UK with output marketed and exports, 1980–1994 (MAFF, 1995a).

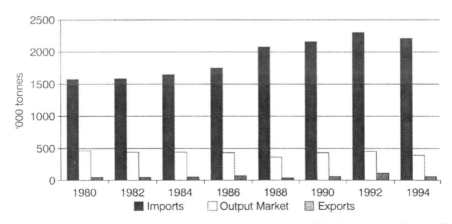

Figure 4.2 Comparison of the quantity of fruit imported into the UK with output marketed and exports, 1980–1994 (MAFF, 1995a).

1. The term horticulture is used here to describe the cultivation of most fruit and vegetables.

argue that it is a mistake for a country like the UK to devote scarce resources to enhancing the competitiveness of horticultural products when those resources could earn a much higher return elsewhere. However, if we recognize nowadays that comparative advantage is not based solely on a country's chance endowment with natural resources but can be enhanced by efforts to improve Porter's diamond characteristics and their interaction (e.g. through efforts to upgrade the R&D or other infrastructure) or the other actors in the food chain (e.g. through the efforts of growers and retailers to upgrade quality), it appears less necessary to take a fatalistic view of the industry.

The UK government (through the Ministry of Agriculture, Fisheries and Food, MAFF) is concerned about the size of the UK 'food trade gap' (the amount by which food imports exceed exports, which stands at some £6 billion) and has recognized that fruit and vegetables are important contributors to the shortfall. Combined imports at around £1.6 billion far exceed exports of around £0.3 billion. The question they raise is whether anything 'sensible' be done about this? Sensible in this context means taking into consideration issues like climate and market trends and only acting in a way that is not wasteful of resources.

Concerns about the food trade gap and the significant part contributed to it by horticultural products coincided with the initiation of a national Technology Foresight exercise in 1995 with the aim to bring industry and academe together to understand what technology could do for the quality of life and wealth creation in the UK and to set priorities for research and its assimilation into beneficial practice.

The national Foresight work considered large sectors of the economy and broad areas of science: the food industry was one (HMSO, 1994a), agriculture another (HMSO, 1994b). MAFF decided that it would be useful to have a more targeted and detailed analysis of certain sectors with a view to orienting some of its own research spending. In the first phase it chose three horticultural products, apples, mushrooms and strawberries, each of which can be successfully grown in the UK, on which MAFF spends substantial sums on R&D, but for which imports substantially exceed exports (which are negligible). The objective of the exercise was to identify the potential for market-relevant technological innovation over a time-span of 10–15 years, and to determine to what extent MAFF could, through judicious targeting of its R&D funds, contribute to the UK industry reaping the benefits. The methodological procedure for each of the three exercises[2] (which were undertaken separately in 1995) involved:

- a literature search
- development of a semi-structured questionnaire

2. Ashbourne Biosciences, under the Directorship of Professor Renton Righelato, was the lead contractor for the work. The marketing elements of the study were sub-contracted to Professor Bruce Traill.

- a series of interviews with actors throughout the chain (growers, processors, retail buyers, traders) and with the research establishment
- development of a delphi-type survey instrument[3]
- a one day workshop attended by around 30 industry, R&D and MAFF representatives at which the main findings of the interviews and survey were presented and debated with a view to obtaining, if possible, a consensus on the way forward
- preparation of a final report and recommendations.

The analysis was not carried out specifically using a Porter approach. Nevertheless, the philosophy behind the study was very much along Porter lines and could easily be ordered into a Porter diamond framework. For example, all sectors of the three horticultural product chains were analysed in recognition of the potential importance of related and supporting industries to competitiveness; the need for market relevance meant that special attention was paid to demand changes and the role of retailers; the main objective of the work was to advise government on its allocation of research funds and the organization of the R&D infrastructure, so specialized factors (in Porter terms), particularly those related to R&D, were highlighted; another objective was to advise growers' representatives on the organization of marketing, so firm strategy, structure and rivalry was considered, mainly in terms of whether linkages in the chain were efficient and also whether the industry was too fragmented (the reverse of the usual Porter concern). The existence of such a detailed assessment of the elements of competitiveness in these three horticultural sectors provides an opportunity to look back and reflect on the extent to which a Porter analysis (plus extensions discussed in Chapter 1) can add insights that can guide policy decisions. That is the purpose of this chapter. Following a short description of some of the changes that are influencing developments across the fruit and vegetable sectors, a more detailed analysis of the apple, strawberry and mushroom sectors is undertaken. We then return to the question: can Porter's framework be used to assess reasons for lack of competitiveness and suggest actions to enhance competitiveness?

It has to be acknowledged that the nature of the research method (confidential interviews, analysis of qualitative questionnaire replies and a consensus workshop) means that many of the statements made in this chapter cannot be formally referenced in the manner of traditional academic research. We hope that the reader will take it on trust that apparently unsubstantiated statements in the text reflect the predominant view of a cross-section of actors intimately involved with the industries studied.

---

3. This was in each case a document of some 15 pages which included a series of statements about likely developments in the industry (market and technological) over the coming 10–15 years and which enabled respondents to agree or disagree and comment on each of the statements. This was mailed to around 80 'active' industry participants, again representing all sectors.

4.2   The fruit and vegetable industry

4.2.1   Production

Fruit and vegetables are characterized by the need for intensive culture and perishability when fresh. About half of UK fruit and a majority of vegetables are produced under protected conditions (in greenhouses, under polythene, etc.) and harvesting of both fruit and vegetable crops is still often carried out by hand.

While most vegetables are annual crops, most fruit crops are perennial, which has implications for investment. Apples, for instance, have an economic life-span of 20–30 years with up to 5 years before the trees begin to bear. This represents a high commitment of funds and involves high risk, the more so with intensive systems which are now being practised. The high annual labour requirement in 'top fruit' production (apples, pears, etc.) is inescapable because of the demands for pruning and harvesting. Much soft fruit (raspberries, strawberries, currants, etc.) is cropped for 3–4 years and strawberries have an initial non-bearing period of 1 year. Mushrooms, by contrast, are cropped several times a year.

The UK produces two-thirds of its fresh vegetable requirements but only a fifth of the total fruit consumed. Imports are sourced increasingly from the

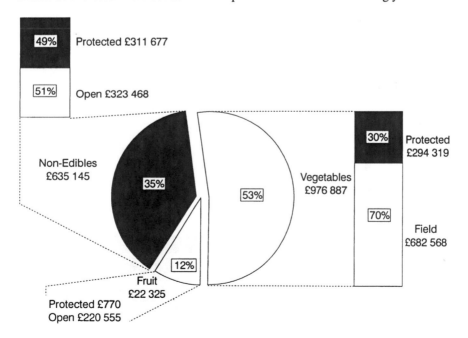

Figure 4.3 Estimated value of UK output of horticultural produce marketed in the UK at current prices, 1994 (MAFF, 1995b).

EU which provides around three-quarters of all the UK's vegetable imports and about two-thirds of fruit.

4.2.2   Consumption of fruit and vegetables and its main drivers

Consumption of fresh vegetables and, more significantly, fruit in the UK is very low by EU standards (see Figure 4.4). Yet there are a number of important drivers for increased consumption (as discussed in more detail in Chapter 2), including:

- health recommendations (for example, COMA, 1994 recommends increasing fresh fruit and vegetable consumption)
- trends towards snacking and grazing (bananas and, to a lesser extent apples, are treated as healthy snack foods in competition with crisps and confectionery)
- the demand for greater variety (it is common to find many varieties of produce on retailers' shelves and, in common with food markets in general, there is an important trend to a greater range of products and a greater turnover of products (product life-spans are shortening)
- improved technology in storage and chilled distribution that ensure better and more consistent quality
- extension of availability of fruit and vegetable products throughout the year through imports and/or domestic season extension has diminished seasonality of many fruit and vegetables
- convenience (ready prepared fresh fruit and vegetables)
- a general consumer trend towards higher quality, usually associated with higher price

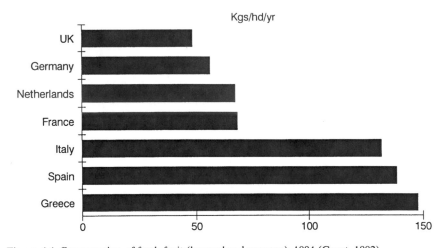

Figure 4.4 Consumption of fresh fruit (kg per head per year), 1994 (Geest, 1992).

4.2.3   The supply chain

Traditionally, fruit and vegetables were sold by specialist greengrocers who purchased from wholesalers who in turn purchased from individual farmers. The rise of the multiple retailer has to a very large extent eliminated this supply route. The retailers prefer to buy direct from growers, bypassing wholesalers, but they prefer to buy bulk supplies from a relatively small number of suppliers. They have thus been instrumental in bringing growers together either as full grower co-operatives or, at a minimum, to collaborate in the marketing of their produce.

Of all UK food and drink sectors, the fruit and vegetable industry is the least involved in processing. Fruit and vegetable consumption in the UK is predominantly in fresh rather than processed form. However, prepared and processed fruit and vegetables are a significant growth area in the market. Processing may involve cutting/slicing, cleaning, packing, juicing, pulping, canning or freezing, and the final product may compete directly on the supermarket shelf with fresh products.

The processing sector is far more concentrated than the fresh produce sector and branding is common in processed products, unlike fresh produce. Nevertheless, the sale of processed produce under retailers' private-label brands has increased and resulted in greater price competition and margin pressure in this sector. Many of the large shippers and processors sell into supermarkets under private-label brands. Also contributing to price pressure is the fact that many processed products are more or less 'commodities' in which quality is relatively homogeneous and competition therefore based on price. Examples are concentrated fruit juice and fruit pulp. Being less perishable than fresh produce, such commodities often come from low-cost producers abroad and in this context the removal of trade barriers (through GATT and CAP reform) and the opening up of the eastern and central European countries is seen by some as posing a significant threat to the domestic industry.

The concentration of demand in the big retail chains has resulted in only large companies with an international strategy being able to market branded products. The products are distinguished not only by their high quality but also by their broad availability in a wide range of retailers and countries. This enables the margins to be maintained without prices becoming appreciably higher than private labels. Brand names are especially important in the canned fruits segments.

Within the fruit and vegetable processing industry a breakdown by value indicates that canned vegetables account for 40%, frozen vegetables 59% and dried vegetables less than one per cent of total vegetable processing. 70% of fruit consumed in a processed form is in cans with the remainder dried. There is relatively little freezing of fruit (Tables 4.1 and 4.2).

Table 4.1 Consumption of processed
vegetables in the UK, 1991–1993

Product	£M
Frozen peas	150.0
Mixed vegetables	64.0
Frozen beans	46.0
All other frozen vegetables	147.0
Total frozen vegetables	407.0
Canned tomatoes	83.7
Canned peas	83.3
Canned sweet corn	44.3
Other canned vegetables	62.7
Total canned vegetables	274.0
Dried vegetables	4.6
Total processed vegetables	685.6

Source: Datamonitor.

Table 4.2 Consumption of processed
fruits in the UK, 1991–1993

Product	£M
Peaches	50.0
Pineapples	41.8
Pears	24.3
Fruit Salad	31.7
Mandarins	14.5
Grapefruits	10.8
Other	21.6
Total Canned Fruit	194.7
Sultanas	32.0
Currants	12.7
Raisins	22.0
Others	20.7
Total Dried Fruit	87.4
Total Frozen Fruit	11.0
Total Processed Fruit	293.1

Source: Datamonitor.

4.2.4   EU fruit and vegetable policy

Fruit and vegetables have been less protected than many agricultural products in the EU, but nevertheless EU policy has an important, if diminishing,
effect. For a complete description and for the recent reforms, see Swinbank
and Ritson (1995). The important elements of the policy include:

● **Quality standards** which apply to some 30 different fresh fruit and vegetables. They define grading and labelling requirements and are applicable at all stages of distribution from grower's packhouse to the retail

shop and to imports and exports, but exclude farm sales and produce destined for processing. The aim is to keep products of unsatisfactory quality off the market, to guide production to meet consumers' requirements and to facilitate trade based on fair competition and thereby help to improve profitability. Quality standards, which cover quality, size, packaging and labelling are applied as follows:
- **Extra class:** Excellent quality and usually only special selected produce
- **Class I:** Good quality produce with no important defects
- **Class II:** Produce of marketable quality which is not up to the standards of higher quality
- a **withdrawal system** to provide a safety net in years of poor quality production or overproduction and to stabilize prices in a market characterized by perishable produce and large price fluctuations. It takes the form of compensating producers from Community funds for withdrawing produce of marketable quality from the market, after which it can be distributed free to charitable organizations, hospitals, schools, etc., where this does not adversely affect the market. Although in some countries, some products have reportedly been routinely grown for the sole purpose of being 'withdrawn', none of the three products considered here has seen significant withdrawal in the UK. Small quantities of apples have been withdrawn.
- a complicated system of **minimum import prices** which have restricted imports from some countries outside the EU at certain times of the year (the EU harvest period).

4.2.5 Research in horticulture in the UK

The UK horticultural industry is supported by a research establishment with an international reputation, Horticulture Research International (HRI). It additionally stands to benefit from the work of the research institutes of the Biotechnology and Biological Sciences Research Council (BBSRC) which carries out underpinning research relevant to a range of biology-based agricultural and food industries, with occasional projects specifically of a horticultural nature. Certain universities also have research teams working on aspects of horticulture. MAFF is the largest funder of horticultural research, though part of the HRI budget comes from BBSRC and in some industries this is supplemented by grower and processor funded work (growers generally pay through levies, often administered by the Horticultural Development Council, HDC), often with MAFF contributing. In some sectors (e.g. apples) EU funding is becoming significant.

Some have claimed that the UK industry is better at fundamental research than at technology transfer and that research is supply or product driven rather than market driven.

4.3   Apples

Apples are a small, but important, UK crop with a farmgate value of £100–130 million, but with net imports at twice that value. UK growers can produce good quality fruit which is liked by British consumers, but Cox, the variety on which much of the UK industry depends, has suffered serious quality problems which have cut prices and put customer loyalty in jeopardy. On the other hand, there is potential scope for growth in the total market, and in UK producers' share. Neither should export expansion be ruled out. However, if these goals are to be realized, a number of constraints must be overcome:

- apples are consumed all year round but the Cox growing season is very short
- quality deteriorates in storage after a few months
- consumers and retailers want variety.

This section analyses the competitive prospects of two distinct sectors of the apple industry, fresh consumption, for which Cox is the main UK variety, and 'culinary' (cooking and processing apples) for which Bramley is the main UK variety, though Cox and other fresh apple varieties can also be used for cooking.

4.3.1   Background to the study

Traditionally apples were produced and consumed in the same locality, but early in the twentieth century the UK began importing fresh apples from southern hemisphere Commonwealth countries such as New Zealand and South Africa.

Around 50% of apples are now sold through multiple retailers, a near doubling in the last 10 years. This is in line with the figures for all fruit and vegetables. The multiples set the quality and variety trends for the market as a whole. Multiples buy directly from grower representatives at home and abroad. Only greengrocers, independents and farm shops use traditional wholesale markets. Supplies are bought by the small caterer from retail outlets, whilst the larger caterers buy from wholesale markets, though it is predicted that catering will also move more towards direct sourcing.

World-wide, the apple market has 'globalized' and 10% of production is traded internationally (see Rabobank, 1992a, 1993a,b for a more detailed discussion), but everywhere varieties with only local appeal can be found alongside the international 'commodity' varieties (Granny Smith, Golden and Red Delicious). The UK seems to have taken globalization more seriously than other European countries, as shown by Table 4.3. She has substantially lower self-sufficiency than other EU countries, even those with similar climates.

If consumption were particularly high, it might be argued that production simply could not keep pace. However, as Figure 4.5 indicates, this is patent-

Table 4.3 Self-sufficiency in apples, 1989–1990

Country	Production (thousands of tonnes)	Consumption (thousands of tonnes)	Production/ Consumption (%)
Belgium/Lux.	297	290	102
Denmark	92	138	67
France	1852	1233	150
West Germany	1762	2323	76
Greece	280	195	144
Italy	2081	1822	114
Netherlands	375	364	103
Spain	686	801	86
UK	342	767	45

Source: O'Rourke, 1994.

ly not the case – consumption *per capita* is the lowest in Europe (Figure 4.5). The view that surely something can be done about this situation forms the background to the original study.

4.3.2 Demand conditions

Fresh dessert apple consumption has been relatively stable at its low level over many years. By contrast, culinary apples have shown a marked decline in fresh consumption rates, estimated at 3% per year. This decline is long term and reflects both an ageing consumer profile and changing lifestyle habits throughout the population. Of most significance is the decline in traditional cooking and eating habits, such as home baking. Retailers estimate that most Bramley consumption is among over 45 year olds.

Apple consumption in the UK is fairly evenly spread through the year. Almost 50% of households purchased apples in the National Food Survey's survey week (MAFF, 1995b), a higher figure than for any other fruit. Retailers believe that there is strong consumer loyalty to Cox and other traditional 'English' varieties, but that loyalty to UK grown apples could be lost

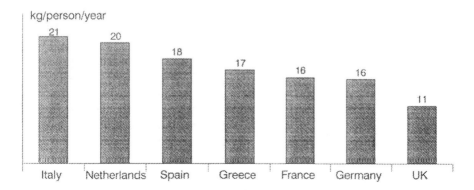

Figure 4.5 Consumption of fresh apples (Geest, 1992).

by recurrent episodes of poor quality Cox, and the preference for 'English' may decrease as the UK becomes more European. Some supermarkets have started to reintroduce 'traditional' English varieties on an occasional basis.

Innovation has become a feature of what historically was considered to be a conservative market. New, high quality varieties such as Braeburn, Fuji and Gala now sit alongside Cox, Granny Smith and Golden Delicious. Supermarkets and greengrocers often have eight or nine varieties on sale, sourced from France, the USA, New Zealand, South Africa and the Netherlands as well as the UK. Advances in consumer analysis may help to identify gaps in the supply of UK grown apples with preferred characteristics (e.g. fresh, acid apples) or combinations of characters that might be sought after in breeding and selection programmes. It could also help define the key expectations that consumers have of apples and so could be used to guide apple promotion schemes. The results of an Institute of Food Research study (McFie, personal communication, 1995) support the view that the main UK grown apples fit only one segment of a market with diverse preferences. Far more needs to be known about preferences throughout the year, in different parts of the country, and about the influence of the range of varieties, price and presentation on actual consumer choice.

In summary, looking at the fresh market:

- there are segments of consumers not catered for by current UK varieties
- there are times of the year when UK varieties are not available even if demanded
- there are segments of consumers who have a demand for variety (do not want to eat the same apples all the time).

UK supermarkets have catered for these wishes by importing a range of foreign products; to the retailers, country of origin is unimportant unless it is perceived as important by consumers – and there is no evidence in the UK that this is so. By contrast, foreign retailers still sell almost exclusively domestic apples (and at relatively low prices).

Does this mean that UK consumers are relatively sophisticated and that this forms a basis for competitiveness? Perhaps, but clearly to date this has not been translated into an advantage for UK growers. Quite the reverse: sophisticated UK retailers have sourced from abroad while less developed foreign retailers have sourced domestically. Can it be an advantage in the future? To the extent that British growers (or their agents) have become adept at dealing with the high quality requirements of UK retailers and to the extent that foreign retailing will move down the same path as British retailing, this should provide opportunities to supply high quality differentiated products (e.g. UK varieties) to foreign retailers. However, there is also the possibility that foreign growers who have similarly become adept at supplying UK supermarkets will similarly be able to take advantage of growing consumer and retailer sophistication in their own and other countries. There is nothing really to suggest that UK producers will be able to appropriate any of the benefits of sophisticated UK demand.

Although the fresh sector is still the most important both by value and output, processing and juicing together constitute a significant and growing sector of the apple industry. Processing apples for later incorporation into finished food, other than to juice and cider, is reported in government statistics to be around 2% of UK apple supplies. In other EU countries processing accounts for between 10% and 25% of production and in the USA, 40%. Consumption of apple juice at around 115 million litres per year is relatively stable and represents around 10% of all juice consumption. Fresh (40 million litres) and carbonated (60 million litres) juices are the growth areas, and the UK exports these products.

Almost all countries other than the UK have 'lost' their specialist culinary apples and use eating apples for cooking and processing. There has been growing interest in the UK in developing Bramley apples as a widely used food ingredient for both savoury and sweet dishes. While the bakery sector is increasingly using Bramley as a brand product, new product development has seen the inclusion of this apple variety in a wide range of products such as sauces, meat and pasta dishes. Both Bramley and Cox also have potential in higher value juices. There is the potential for likewise developing exports of this unique 'brand'. In the Porter context, a sophisticated demand (in fact, a traditional demand that has not been lost) could provide a source of international competitive advantage.

4.3.3   Factors of production

Growing conditions.   As already indicated, Cox represents two-thirds of eating apple production in the UK. Bramley is even more dominant in culinary production, with 92%. (See Figure 4.6.)

Large variations exist between growers in the quantity and quality of yields. This is mostly attributable to variations in the standards of husbandry in the industry. There is a controversy as to whether more intensive systems require higher standards of husbandry than extensive systems. However, both approaches require expert husbandry skills to maximize returns and both systems can be productive and profitable.

UK average yields (around 18 tonnes per hectare) are low compared with those of major competitors (O'Rourke, 1994). The main reasons reported for the lower average yields in the UK are

- dominance of lower yielding varieties, mostly Cox
- relatively short season and frequent cold spring periods.

However, more advanced and intensive growers regularly exceed 30 tonnes per hectare of Cox and some other varieties that grow in the UK, such

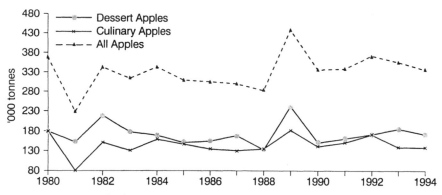

Figure 4.6 Gross production of apples in the UK, 1980–1994 (MAFF, 1995a).

as Fiesta and Royal Gala, have yielded over 50 tonnes per hectare in recent trials.

The UK climate is in some respects an obstacle to apple producers, but not an insurmountable one, and in other respects it is even an advantage. The climate is thought to account for some of the colour and flavour advantages of English apples. For example, Royal Gala grown in the UK are claimed by retailers and growers to have better colour and taste characteristics than those grown in France; Cox when grown in the Netherlands or New Zealand has poorer taste and texture.

The risks associated with Cox can be reduced and margins improved by a variety of means already adopted by some growers. The most critical issues are maintenance and prediction of quality in storage, and fruit size and quality. Adoption of the better Cox clones and appropriate crop management strategies are reported to optimize the latter of these.

There exists a considerable lag between planting and first harvest of apples and, despite advances in technology, it is 5 years before an orchard reaches full bearing. Establishment costs are £5–10 000 per hectare (planting plus maintenance). To maintain competitive production of high quality fruit, orchards usually have to be replaced after 15–20 years. (Strathclyde University Food Project, 1993, 1994).

Average gross margins of £2500 per hectare, rising to four times that figure for new high-intensity orchards with premium varieties, could generate acceptable returns, but the risks associated with long paybacks in a changing market, and the uncertainty of quality and yield, do not make dessert apples an attractive investment to many growers.

A number of factors could reduce risk:

• growers holding a portfolio of varieties
• co-ordinated planting, production and marketing of new varieties
• reduced time from planting to producing a marketable crop.

Storage and packing.   As consumer expectations of quality continue to rise, and developments in analytical and information technology permit more and more information about a product's quality and history to be gathered, the demands on producers to assure more aspects of quality will grow and present a source of competitive advantage. This will require more sophisticated storage, grading and packing to meet the standards of retailers and processing markets which will encourage co-operative storage, grading and packaging operations. The industry requires fewer packhouses but a better geographical spread and the capacity to deal with large volumes, since volume capacity is of great importance to retail buyers looking to source large volumes from a single outlet.

Research.   The level of research expenditure, around 3% of farmgate output value, government support at around 80% of this, and the gearing of industry research funds with government support are all probably higher than for any other agri-food sector, and the opportunity for the industry to influence research is disproportionate to its size. Current main objectives of HRI programme include fruit quality, precocity and disease resistance in the red and Cox-like apples. In, 1994, following calls from the industry, greater emphasis was placed on fruit quality parameters.

The technology for transforming (genetically manipulating) apples exists and the capability has been established at HRI. The range of genes available to alter properties important in apples is increasing rapidly, e.g. genes controlling ethylene synthesis, genes blocking pectin breakdown, and a range of antifungal and insecticidal proteins. Genetic transformation in principle provides the opportunity to design into apples many of the features required by consumers (colour, texture, sweetness, acidity, flavour) and for orchard management, such as texture stability, fungal disease resistance and non-browning.

Because of the long time-scale and the relatively small value of the market for apple stock, apple transformation is not a priority for commercial plant genetic companies and without government support the national and international programmes would not be established.

New non-invasive sensing technology is being developed for environmental monitoring, health screening and many aspects of food quality, and will be available for adaptation to apple grading and apple product testing.

In summary, as far as the quantity of R&D is concerned, the industry seems well supplied in relation to other agri-food sectors and probably in relation to the industry in other countries. There has been some criticism that in the past the direction of research has been too much supply driven and that HRI has developed few (if any) new varieties that have become commercially accepted. The research undertaken by HRI is leading-edge work internationally, but to what extent can this benefit UK producers as opposed to the international community as a whole? In principle, work directed at Cox or

Bramley can benefit UK producers, but consumers are moving away from Cox in particular. If new varieties were developed that gave the UK industry a first-mover advantage this would also benefit the UK grower, as would adaptation of foreign varieties to suit them for the UK climate and the development of varieties particularly favoured by UK consumers.

4.3.4 Related and supporting industries and firm structure, strategy and rivalry

Marketing has become increasingly concentrated as growers have responded to the multiple retailers by forming larger co-operatives. There are now eight major selling desks, the largest trader having around 50% of the dessert and culinary market. This is a very high level when compared to the rest of British agriculture, but remains a low figure when compared with levels of co-operation in some other apple producing countries. British growers have also learnt to work flexibly and closely with British retail multiples, the largest sector of the quality fresh fruit market. Concentration provides better opportunities to manage quality at harvest, to optimize storage, control the shares of Class I and II produce entering the market, and introduce new varieties.

It is not apparent that there are fundamental barriers to British competitiveness arising from the structure and interconnectedness of the food chain. While some have pointed to the benefits of the Dutch 'auction' system or the New Zealand export marketing board, it would seem that the UK industry has modernized its marketing system substantially in recent years.

4.3.5 The basis of competitiveness

British apple growers cannot compete with other EU member states in producing apples (e.g. Golden Delicious) for commodity markets because output per hectare in warmer climates is several times the yield that can be achieved in the UK. Apple varieties grown in the UK need to reflect the prevailing climatic conditions, the slower maturing of fruit and the more distinctive flavour. However, production of new full flavour varieties, such as Braeburn and Royal Gala, in countries such as Germany and New Zealand, remains a threat to the British industry.

There is no single over-riding reason for the extent of the UK's current lack of competitiveness in apples. It would seem to be a combination of a number of factors, some of them set far back in history but, given the scale of the time lags for this crop, still having their influence (research started now would require at least 12 years to develop a new variety with significant market presence). Demands by retailers for variety have had to come from imports; a poor selling structure in the past meant inconsistent quality for Cox and meant that retailers wanting large quantities of uniform (high) quality produce would best look abroad. Perhaps also the research establishment, under pressure to produce short-term results, concentrated on marginal improvements in disease control, storage and in Cox quality at the expense of more innovative work.

Overexposure in the market to Cox, its low yields and variable quality argues for the introduction of new varieties to give a more balanced portfolio of dessert apples. Current breeding programmes still focus largely on Cox-like varieties. For mid-term introductions (5–10 years), systematic screening of varieties from the breeding programmes of other producer countries as well as from the UK programme could increase the chance of market success, as could a greater focus on finding out what characteristics UK consumers specifically want and breeding varieties with those characteristics.

4.4 Strawberries

The situation for strawberries is in many ways similar to that for apples:

- a seasonal product with traditionally a very short consumption period, but imports by retailers allowing all-year availability in recent years
- a climatic disadvantage in terms of yield, but some advantage for the domestic growers in that climate and/or freshness of the domestic product improves flavour
- a strong R&D system, again provided by HRI
- considerable reliance on a single variety (Elsanta), but a consumer demand for variety.

4.4.1   Background to the study

Strawberries are one of the largest horticultural crops in the UK, with a farmgate value of around £60 million and total retail sales including imports of £150 million. The product is usually profitable for growers despite climate differences from year to year.

The UK is about 55% self-sufficient in strawberries over the year, with UK growers having over 80% of the market in June/July, during which about 86% of the UK crop is sold (Figure 4.7). Maincrop production, largely Elsanta, is harvested in June and July and represents around 80% of the 42 000 tonnes of strawberries produced in the UK. Over the period 1980–1993, total cropping area has declined by 34% from 7920 hectares to 5192 hectares. During the same period, gross yields have increased by 22%. Gross yields per hectare in the UK rarely reach 10 tonnes per hectare, 5–8 tonnes per hectare being more normal for a good grower, compared with yields around three times higher in southern Europe and California which have higher light intensity, heat and longer production seasons. Overall, gross production has declined by 21% from 53 400 tonnes in 1980–1981 to 42 400 tonnes in 1994–1995 (Figure 4.8).

The yields and costs of strawberries vary through the year and with the many different types of growing systems. However, with over half of costs

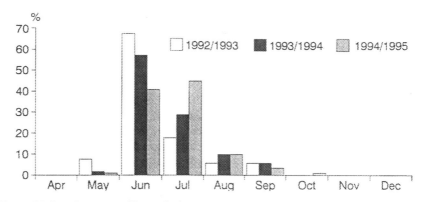

Figure 4.7 Strawberry monthly marketing pattern: proportion of the output marketed in the given month (MAFF, 1995a).

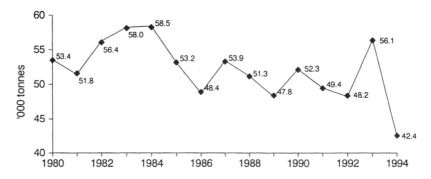

Figure 4.8 Gross production of strawberries in the UK, 1980–1994 (MAFF, 1995a).

associated with picking, packing and transport, yield per hectare, though important, is not an overwhelming factor in competitiveness. UK growers dominate the home market in June and July, and some can compete with American imports later in the year.

Figure 4.9 gives an overview of the main distribution channels for fresh strawberries in the UK. It shows that imports represent a little under 50% of consumption.[4] Of the UK produced strawberries, almost equal amounts are sold through multiple retailers, pick-your-own[5] (which is in long-run decline) and 'other' outlets (farm shops, greengrocers, etc.). Consumption outside the home is around 20%. The share of UK sales going through the multiple retailers is considerably lower than for most fresh produce at around

4. Industry production figures differ from those presented in Figure 4.9 because the industry claims that official statistics overstate the UK crop.
5. With pick-your-own, growers permit consumers to pick their own strawberries and charge according to the quantity picked. Lower prices and yields are compensated for by the absence of picking costs.

Fresh Strawberry
Distribution Channels

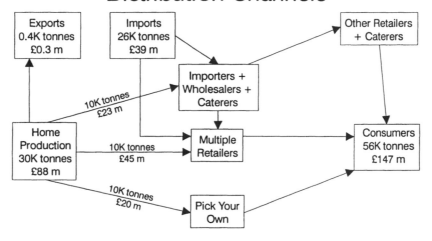

Figure 4.9 The distribution channel of fresh strawberries (industry estimates).

30–40% by volume, probably because of limited shelf-life combined with once a week shopping from supermarkets. This could represent a long term constraint on consumption growth.

Imports have grown rapidly (Figure 4.10) and spring strawberries coming from southern Europe, particularly Spain, compete with the early UK produce. Supplies of strawberries after the UK summer season, from California, South Africa, South America and India have developed a new market which UK suppliers are now entering (Table 4.4).

Prices, particularly out of season, are determined by European (and to an extent, global) supply and demand conditions. Increased penetration of the German market by eastern European suppliers could displace Dutch and Belgian supplies, which would normally go to Germany, on to the UK

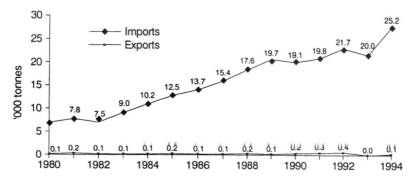

Figure 4.10 Import and exports of strawberries into the UK (MAFF, 1995a).

Table 4.4 UK strawberry imports (12 months to December 1994)

Imports	Intra-EU		Extra-EU		Total	
	Tonnes	£000	Tonnes	£000	Tonnes	£000
1 May–31 July	12 502	17 153	1 150	1 144	13 652	18 297
1 August–30 April	9 340	15 610	2 825	4 853	12 165	20 463
Total	21 842	32 763	3 975	5 997	25 817	3 860

Source: C.S.O. Business Monitor Overseas Trade Statistics, 1994.

market, depressing prices in the mid term (5–10 years). The premium for out of season UK strawberries is narrowing and likely to narrow further under Californian, southern hemisphere and southern European pressure.

The potential opportunities for UK growers are therefore

- to substitute some of the net imports of around 26 000 tonnes by extending the growing season
- to develop export markets
- to increase consumption of home grown strawberries.

The threat comes from improved quality and lower priced imports (Strathclyde University Food Propject, 1992).

4.4.2 Demand conditions

The consumption of soft fruit (mainly strawberries) has fallen slowly over the last two decades, but strawberries are still a highly preferred, luxury fruit.

There is a markedly seasonal pattern to fresh strawberry consumption (Figure 4.11). This is despite year-round availability in supermarkets, though it must be admitted that winter supplies are expensive and tend to have less flavour. Good weather within the peak period creates surges in consumption. The seasonal consumption pattern is likely to become less marked as better quality, cheaper fruit becomes available from California, South America, India and elsewhere. Data are not separately available for catering, but it is probable that a significant share of strawberry consumption is outside the home.

Consumption data are not readily available for other countries, but, in common with other fruit, it is likely that UK per capita consumption is below that of southern Europe (figures as high as 16 kg per capita have been quoted to us for Spain, compared to less than 2 kg per capita in the UK). It has been reported that even in the peak consumption month of June, only one in three UK households purchased strawberries in any one week.

Retailers are seeking to differentiate their products from those of their competitors by exciting customers with the different tastes of a range of labelled varieties. However, this process has not yet proceeded very far, as

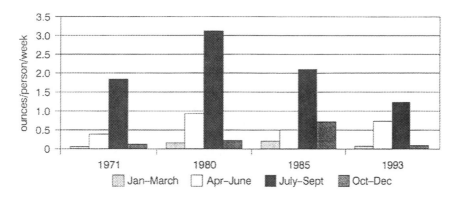

Figure 4.11 Seasonal consumption pattern of soft fruit (other than grapes) in the UK (MAFF, 1995b).

strawberries are rarely sold by variety in the way that apples are. Supermarkets have tended to buy Elsanta in preference to others because of the consistency of its quality. Although other varieties may have superior taste they are not as firm as Elsanta, which has now become the standard against which other varieties must compete. Current specifications for strawberries may include size, sweetness and colour, but flavour characteristics are not defined and there are few consumer preference data on which to base development of differentiated products. Strawberries in future may be made more interesting to retailers by increasing the product line to include wild strawberries, large ones for dipping, designer supersweet strawberries, fresh strawberry sauces, snack-packs for the school or office, breakfast strawberries and dehulled strawberries.

British strawberries are preferred to Spanish and other imported strawberries by consumers and retailers because of their flavour. This is closely linked to freshness. A key question for the UK industry is whether to aim for increasing shelf-life with a view to increase consumption (particularly through multiples). This strategy has an inherent risk that foreign producers can match UK strawberries for freshness and hence flavour. A similar conundrum exists with regard to yield enhancement. Will higher yield reduce flavour and hence dilute the perceived advantage of UK strawberries? Certainly, the perceived quality advantages of the home-produced product are the greatest asset the industry possesses and any new variety development must respect this advantage.

In conclusion, a sophisticated retail demand has led to increased domestic consumption but this has tended to benefit foreign rather than UK producers. The UK has not been able to translate its expertise in satisfying the needs of sophisticated buyers into a positive trade performance.

4.4.3   Factors of production

Growing conditions.   High quality is an essential element of the competitive advantage of the UK supplier. For a fragile, short-life product, both quality and cost benefit from proximity to the market. Proximity to the home market provides a quality advantage to the UK grower, giving about 24 hours less time from picking to retail store. This allows a better eating quality product to be offered, and a longer shelf-life. It is also claimed that Elsanta has better flavour than imported varieties, but this may be because the fruit is fresher. However, with better organization of the chill chain, improved packaging techniques and reductions in the cost of transport in real terms, the advantage of home producers is being eroded.

Research.   Molecular genetic transformation is beginning to affect strawberry breeding. Transformation systems have been developed for strawberries and transgenic plants have been produced for experimental purposes since 1990. It was reported that the gene for the protein sweetener thaumatin had been put into a strawberry in California. The UK has a strong science base in plant biotechnology, and HRI has a leading programme, in world terms, in strawberry research. It is expected that genetic transformation will be a major element of future breeding programmes, which will speed up the generation of new varieties and generate step changes in agronomic and quality traits.

The industry is concerned that current levels of research and development are not adequate to take advantages of opportunities. Low profitability in the industry means that it is difficult for it to fund substantial levels of R&D itself, and cutbacks in government funded research have exacerbated this situation. Research priorities include new varieties and species to complement Elsanta in other market segments and for the extension of the growing season.

As with many other horticultural crops, the small market size in the UK for the strawberry plants provides only a small direct return to breeders. Investment of public funds to overcome the resultant dysfunction in development can only be justified if it is clearly linked to much larger benefits further along the value-added chain. Greater involvement of marketing organizations and retailers in the breeding programmes could be sought, with the potential for exclusive arrangements and royalties linked to sales.

The linkage between growers and the R&D programmes is thought generally to be good and there is a strong input from some growers into the HRI breeding programme. It has been suggested that this could be improved by reaching a wider audience of growers around the country and by using objective, quantitative measures of quality linked to consumer research.

4.4.4   Related and supporting industries and firm structure, strategy and rivalry

Strawberry growing in the UK is made up of a large number of small producers, over 2000 according to MAFF statistics; 2% of these account for nearly one-third of the total acreage. The trend to larger operations and central packhouses serving groups of growers is expected to continue, driven by capital and management demands and the requirements of supplying the multiple retailers.

In parallel, there has been a reduction in the number of marketing desks, so that now over half of production goes through seven marketing organizations. These trends facilitate the supply to the multiples and could promote the development of new products and entry into export markets.

Almost all UK strawberries are sold as fresh fruit. There is substantial use of strawberries in processed food, perhaps equivalent to around 10–20% of the fresh sales volume. However, most of the processed strawberries used in the UK, in yoghurts, preserves, baking, etc. are imported from lower cost countries, particularly eastern Europe.

With increasing interest in fruit products and the continued growth of freshly processed products for sauces, purées, juices, etc. in the chill chain, the consumption of fresh strawberries in processed food is expected to grow, and could offer opportunities to the UK industry.

4.4.5   The basis of competitiveness

There appears to be substantial scope for growth in consumption of strawberries, through

- greater availability at reasonable prices outside the summer season
- increasing the proportion of people choosing strawberries in the peak season
- increasing the opportunities to eat strawberries by increased shelf-life at home.

The multiple retailers, who predict continuing rapid growth in their sales and who will soon account for over half of sales, are central to the development of the extended season, to quality standards and to adding value through product differentiation.

The rate of introduction of new varieties will increase, driven by demands of the market for higher quality and diversity and, on the supply side by the contribution of molecular genetics and international competition in breeding. However, any attempt to develop new UK varieties must be very careful not to put at jeopardy the perceived quality of the UK grown fruit.

4.5   Mushrooms

The mushroom industry bears some resemblance to the apple and strawberry sectors, but there are some significant differences. The most notable difference is that mushrooms are not seasonal and, being capital and labour rather than land intensive, they do not compete in the supply decisions of farmers in the way that apples and strawberries do. Also significant is that the UK is much more self-sufficient in mushrooms and that the industry is quite concentrated. Similarities are reliance on a single variety at a time when the market is showing signs of product differentiation, the vital role of retailers as the main buyers, research support from HRI and, more importantly than for the two other products, the threat of competition from low-cost producers in the emerging markets of eastern and central Europe.

4.5.1 Background

The edible mushrooms sector comprises the cultivation and processing of, and trade in, edible fruiting bodies of a number of families of fungi. The main representatives of edible mushrooms are: the cultivated mushroom (*Agaricus bisporus*), the oyster mushroom and the shiitake (Mushroom Growers' Association, various)

Mushrooms are the largest horticultural product grown in the UK, with a farmgate value of around £170 million and total retail sales in excess of £250 million. The UK has traditionally been largely self sufficient, though imports, supplying about one third of the market, have taken much of the growth in the market over the last decade.

Producers are confronted with a rapidly rising demand for fresh mushrooms, but that for the preserved product is growing less quickly. A higher demand for fresh mushrooms strengthens the position of producer countries located close to the consumer centres. Because of the perishable nature of fresh mushrooms they are unsuitable for transportation over very long distances (there is virtually no intercontinental trade). World trade is limited to preserved mushrooms (mainly canned or dried).

Figure 4.12 gives an overview of the main distribution channels for UK mushrooms. It shows that of the total annual UK consumption around 140 000 tonnes, just over a third is eaten outside the home and of the remainder, more than 70% is sold through the multiple retailers.

Multiple retailers generally source directly from home growers (particularly the larger ones), or from foreign growers or their agents. Imports of mushrooms into the UK have risen sharply in the space of ten years from 10 900 tonnes in 1984 to 53 900 tonnes in 1994 (Figure 4.13). The Republic of Ireland has been the largest supplier of fresh mushrooms since 1988, accounting for 29 640 tonnes in 1994. This country supplies a very good quality product and responds skilfully to the wishes of the large British supermarkets. Following Ireland, the Netherlands is the second most important source of mushrooms into the UK market. In 1994, exports of Dutch fresh mushrooms to the UK were around 8000 tonnes. Competitiveness is very sensitive to the exchange rate, which has adversely affected the Netherlands and Irish exporting to the UK, but opens up prospects for UK exports (Rabobank, 1992a,b).

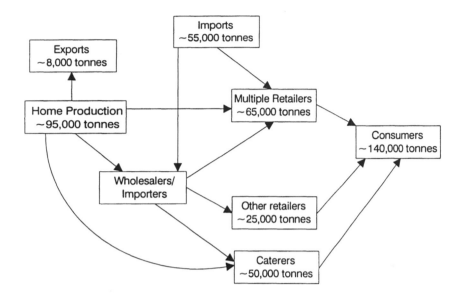

Figure 4.12 The distribution channel of mushrooms (industry estimates).

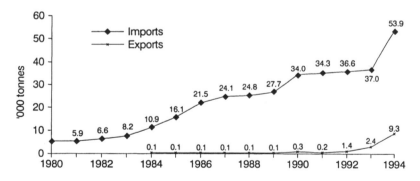

Figure 4.13 Imports and exports of mushrooms into the UK (MAFF, 1995a).

4.5.2 Demand conditions

Unlike most fruit and vegetables, where UK consumption is low relative to other European countries, mushroom consumption is at a comparable level with other Northern European countries and significantly higher than levels in Southern Europe. Mushroom consumption more than doubled during the 1980s to 2.9 kg per head per year (Figure 4.14). The growth in UK

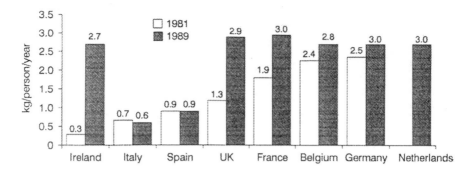

Figure 4.14 Mushroom consumption, 1981–1989 (Geest, 1992).

consumption is thought to have arisen from a large-scale, successful advertising campaign combined with a fall in real prices.

Mushrooms are reported to be purchased on the basis of appearance and price, rather than flavour, eating or cooking characteristics. Mushrooms are not promoted on the basis of country of origin, and there is therefore no apparent consumer loyalty to home grown produce. There has been very little segmentation of the market; 98% of sales are of *Agaricus bisporus*. Although the share going to open flats, chestnuts, etc. has increased, they still represent less than 2% of the total. 'Exotics', mostly the oyster mushroom and shiitake, have around 0.5% of the market. There is a small market for exotics picked from the wild in the UK and imports of fresh and dried 'wild' fungi. Together these probably account for less than 1% of total expenditure on fungi in the UK.

Despite apparent consumer interest in exotic fungi, as evidenced from media attention and displays in shops and restaurants, this has not so far translated into significant consumption. However, it is probable that market segmentation will occur to a greater extent than in the past, presenting opportunities for new product development, e.g. better quality preserved mushrooms (particularly in jars), frozen mushrooms, mushroom salads, distinctly flavoured mushrooms, products for catering and prepared foods.

Retailers have suggested that there are opportunities for more highly flavoured mushrooms and other specialities, but there is little consumer research to direct product development. Moreover, to develop the market for exotics, retailers need to be assured of a stable source of high quality product.

Few UK produced mushrooms are processed, though growth in prepared mushrooms products (e.g. soups, patés, sauces) is expected, driven by the growth in vegetarianism and by newer, milder preservation techniques. Although they are high in fibre, low in saturated fats and contain a number of vitamins, mushrooms have not yet been marketed in the UK for their nutritional properties.

4.5.3 Factors of production

Growing conditions.   There are several production systems, ranging from the integrated composting and shelf or tray growing operations of the larger producers, to small scale growers, many of whom now buy bags or blocks of 'phase II' compost (compost ready prepared and inoculated with spawn). In all systems the basic growing process is the same. Mushrooms are grown on compost produced from manure and straw. The mushroom spores are initially spawned on cereal grains before being mixed with the compost. Mushrooms grow very rapidly, and within a week of spawning the first flush is harvested. Each bed will produce 3–4 flushes. Each square metre of compost produces an average of 22 kg of mushrooms per growing cycle. The average number of growing cycles per year is five.

The bag system has no important economies of scale in production, and has made the smaller grower competitive with larger producers. In fact, small scale production offers advantages because of lower cost labour in family enterprises and the fact that production requires intensive attention to detail which is better suited to small scale activity.

The main cost elements are similar for the different production systems. Compost, spawn and casing amount to around 30% of the costs; labour (mostly picking and packing) adds another 35%; fixed cost represents a further 25% of total cost. Legislation to improve health and safety in the workplace could increase labour costs. The main issue is repetitive strain injury arising from picking, and other injuries from awkward access to beds for picking.

Thus the mushroom sector is labour intensive and over the past 10 years the sector has seen the labour cost go up by virtually a third in some areas. Labour costs, however, are higher in Holland and France but lower in potential competitor countries in eastern Europe, e.g. Poland.

Research.   Growers in the UK fund research, through a levy on spawn paid to the Horticultural Development Council of around £150 000 per year. Almost all the research is carried out at HRI. The programme of research is set by a panel led by growers and deals largely with urgent common problems of disease control, composting, etc. MAFF funded research is currently about four times this level and is intended to underpin industry research by maintaining a strategic research programme for mushroom physiology and disease and supporting the initial stages of novel technology development. HRI recently built a new pilot composting and growing facility. The BBSRC, through its core grant to HRI and through research programmes at universities, supports basic research in fungal physiology and genetics, and in other underpinning areas of biotechnology, pathology, etc. Molecular genetics has been a particular area of investment.

Unlike most horticultural products, the controlled production environment means that mushroom technology is essentially the same world-wide and the research issues are largely common. So, although there may occasionally be opportunities to gain national advantage through nationally owned research, it is the consensus of both growers and scientists that much more is to be gained by international collaboration. It is therefore surprising that there are no major EU programmes on mushroom research.

The technology exists for production of many species in addition to those already grown in the UK, most of it arising from the far East. The know-how for production of the few exotics grown in the UK (oyster and shiitake mushrooms) is jealously guarded by the few producers who supply this limited market. Production system innovation and new product development is usually confidential and unquantified. It relies on acquisition of information from other growers, suppliers, research institutes and leads some growers to achieve yields several times the average for the industry.

4.5.4   *Related and supporting industries and firm structure, strategy and rivalry*

There are three large producers of mushrooms in the UK, which largely determine the image of British mushroom growing: Blue Prince (Heinz); Chesswoods (Tompkin) and Middlebrook (recently sold by Booker to the Irish company Monaghan). Together, these account for a third of production; a further third is from 20 medium sized producers, and the final third is supplied by about 300 small producers. This structure has remained unchanged over the last 20 years.

The supermarkets are gaining an ever-increasing share of mushroom sales: 69.1% of fresh mushrooms sold for home use are now sold through the major retailers who source direct from large (and some medium-sized) producers and importers. The supermarkets set high requirements for the product, particularly concerning the timing of delivery (stores often want to have half of the fresh mushrooms delivered on a Thursday in preparation for the weekend sales). Small growers usually sell through wholesale markets which have become very much residual, with prices determined by the vagaries of supply and demand and sensitive to availability from the Netherlands and Ireland as well as the UK. Therefore, prices are variable and generally substantially below the prices multiple retailers pay to growers. There is limited co-operation in the marketing of mushrooms.

Mushroom growers have attempted to boost demand through generic promotion. In the, 1980s, a promotional campaign by the Mushroom Growers Association was so successful that total demand could not be sourced from UK supplies and the main beneficiaries were foreign growers.

It is relatively easy to enter mushroom production on a small scale; however, lack of production know-how and the difficulties of marketing on a small scale have resulted in a high turnover of new entrants.

Most compost is produced in large units and used on site by the integrated producers, or sold to the smaller and medium sized growers. Mostly this is now phase II compost, prepared and inoculated with spawn. Phase III compost, in which development of the mycelium in the compost is done by the composter, is also available. Phase III produces a crop within 10–20 days, saving on building investments and fixed costs, and is being used by an increasing number of growers. Compost supply is for the moment a domestic and regional industry, but trade is an increasing possibility, as added value moves back along the production chain to the composter.

The need to respond to environmental concerns (mainly odour) is also resulting in a shift in expertise towards composting companies and may result in production at sites remote from residential areas.

Suppliers of equipment, compost, spawn, etc. are important sources of technology for growers and an important route for technology transfer. With the exception of spawn producers and agro-chemicals suppliers, most are small companies, carrying out short term development of their products and drawing on the research institutes in the UK and elsewhere for precompetitive research.

Spawn supplies are largely in the hands of a few non-UK companies selling world-wide. Much of the spawn used in the UK is imported. Spawn production involves relatively high technology, requiring aseptic production, rigorous quality control and a significant research input.

4.5.5   The basis of competitiveness

Mushrooms are grown in a controlled environment and so, unlike most other horticultural products, their production is not subject to peculiarities of geography and climate. Products and production technology are essentially similar throughout Europe and North America. Thus, the competitiveness of the UK grower today is based on the advantages of proximity to the home market and cost of production.

The value of proximity is twofold: time (prime quality mushrooms have to be delivered within 24 hours of picking) and transport cost. UK growers are estimated to have a 15p per kilo advantage over growers from Ireland and Holland. These advantages are decreasing as transport costs fall as a proportion of all costs, and as control of the cold chain improves, facilitating delivery of prime quality product. The factors which reduce the value of proximity to home producers *vis-à-vis* foreign competitors also facilitate exports.

Technology is being developed world-wide, with world-wide applicability. Its impact on UK competitiveness will be determined by the UK grower's ability to adapt and adopt international research and development in advance of competitors. Thus, technology acquisition, training and other aspects of technology transfer are at least as important as home research and development.

4.6 Overall conclusions. Does Porter help?

Three horticultural crops that are important to the UK have been analysed in a pseudo-Porter framework (pseudo in the sense that the studies were not originally planned in a Porter context). The UK has a poor competitive performance in all three in the sense that imports exceed exports and the trend has been towards lower self-sufficiency levels (a loss in market share).

For strawberries and apples, the performance could be put down to adverse climate, but this is simplistic and does not explain why the UK has a lower self-sufficiency rate than countries with similar climates (furthermore, UK demand is low, so a low self-sufficiency rate implies exceptionally low output per head of population). For mushrooms, grown under protected conditions, climate does not matter.

The poor UK performance is not a recent phenomenon, but it has worsened over time: expansion in demand has been satisfied by imports rather than domestic production. For apples and strawberries this can be put down in part to the effects of a general demand trend towards greater choice and diversity. To meet these consumer wishes, retailers have stocked more varieties of apples (a trend which seems to be in the early stages for strawberries and mushrooms as well) and ensured availability of apples and strawberries out of season. For all three crops, the UK is almost entirely dependent on a single variety and producers are therefore unable or unwilling to meet the demand for more variety, let alone off-season demand.

To be successful in the modern food industry where product proliferation and accelerated product turnover is the norm, firms have to be extremely innovative and market oriented. The R&D carried out in the UK at HRI has an international reputation for excellence but has not been successful in turning out new varieties either to suit the specific needs of the UK consumer or that UK producers could use to target a niche-market segment of international consumers. The launching of a new product, particularly one like a new apple with a very long lead time, requires a coordinated effort to ensure that large enough quantities come on line that they can be sold as a segmented 'brand'. This has not happened in the UK.

Does the Porter diamond framework provide a good explanation of the lack of success? If we take the points of the diamond in turn, on the demand side one could argue one positive element, the existence in the UK of perhaps the most sophisticated retail buyers in Europe (or beyond), which Porter would argue should promote upstream competitiveness, and one negative factor, a low aggregate level of home demand. For factors of production, the situation is similar, an excellent research organization associated with high levels of research funding, but on the negative side, generally unfavourable climate (for the two fruits). The other two diamond points were considered together. Historically there has been too little cooperation in the industries to ensure adequate quantity or homogeneous quality to the retailers and this has frequently been blamed for the poor UK performance. Certainly there has been considerable recent consolidation to reduce the number of selling points. Thus lack of co-operation, not lack of competition, appears a serious issue. Related and supporting industries (particularly processors) are not tightly linked into the network for any of the products studied. Many processed products use bulk minimally processed imported ingredients which are traded purely on the basis of price (concentrated fruit juice, fruit pulp) and for which the UK can never hope to compete as an international low-cost producer, but value added processed product niches are opening up and it is not clear that the UK has the infrastructure to react.

Thus the UK has two extremely positive diamond elements and several mildly negative ones. It is hard to take Porter much further. One might argue that the UK problem has been that the diamond elements have not been reinforcing in the way that Porter envisages, so the research output has focused too much on basic research and, at the other extreme, short term issues like disease control rather than communicating with and reacting to what retailers/consumers and growers need (new varieties suited to UK growers and consumers). The lack of communication may result from the lack of co-operation among growers and the absence of related and supporting industries, as well as from retailers' indifference to the source of their purchases (UK or abroad). A fairly compelling case could be made for establishing an infrastructure that encouraged communication among all the actors in the industries (these were in fact among the recommendations made by the studies). However, a fairly compelling case could also be made that globalization in retailing and technology has overtaken the Porter diamond concept. Home base has no real meaning or importance, since sophisticated retail demand is easily translated into imports (and foreign producers are just as adept as UK ones at learning how to deal with UK retailers) and a leading technological edge benefits competitors as much as domestic producers. Readers are free to decide which explanation they prefer.

References

Committee on Medical Aspects of Health (COMA) (1994) *Nutritional Aspects of Cardiovascular Disease,* Report on Health and Social Subjects 46, HMSO, London.
Geest (1992) *Fresh Produce Report* 3.
HMSO (1994a) *Technology Foresight: Progress Through Partnership, Food and Drink.* HMSO, London.
HMSO (1994b) *Technology Foresight: Progress Through Partnership, Agriculture.* HMSO, London.
MAFF (1995a) *Basic Horticultural Statistics for the UK.* HMSO, London.
MAFF (1995b) *The National Food Survey 1994.* HMSO, London.
Mushroom Growers Association (Various issues) *The Mushroom Journal.*
O'Rourke, A.D. (1994) *The World Apple Market.* Food Products Press, Binghampton,NY.
Rabobank (1992a) *International Competitiveness in the Fruit Growing Sector.* Rabobank Netherlands.

Rabobank (1992b) *International Competitiveness in the Mushroom Sector.* Rabobank Netherlands.

Rabobank (1993a) *The World Fresh Fruit Market.* Rabobank Netherlands.

Rabobank (1993b) *The World Processed Fruit Market,* Rabobank Netherlands.

Strathclyde University Food Project (1992) *Opportunities for British Food Suppliers: Action Phase 1992, Soft Fruits.* Strathclyde University, Glasgow, pp. 51–63.

Strathclyde University Food Project (1993) *The British Apple Industry: Market. Opportunities for British Growers* Strathclyde University, Glasgow.

Strathclyde University Food Project (1994) The State of the UK Apple Industry. *Project Paper No. 12,* Strathclyde University, Glasgow.

Swinbank, A. and Ritson, C. (1995) The impact of the GATT agreement on EU fruit and vegetable policy. *Food Policy,* **20**(4).

5 Small firms, old traditions equals low profit: pigmeat processing in Belgium

JACQUES VIAENE AND XAVIER GELLYNCK

5.1 Introduction and objective

Generally, the competitive position of the Belgian meat sector is considered to be worsening. Overcapacity on the one hand and pressure from the retail sector on the other result in fierce competition in the domestic and EU markets.

The objective of this chapter is to identify the competitive position of the Belgian pigmeat sector and its determinant factors on the domestic and export markets. The competitive analysis starts with measuring the profitability of the meat sector. A sector with a high profitability has good competitive potential and can improve its competitive position. The competitive position in external markets is evaluated by applying the refined Boston Consulting Group (BCG) matrix. The Belgian situation is examined for its three main export markets, Germany, Italy and France. The most important components determining its profitability and competitive position are analysed. In this perspective, the competitive factors from the Porter diamond are used (Porter, 1990).

The study is applied to the Belgian pigmeat sector at the level of slaughterhouses and cutting units. In Belgium, meat companies are mostly family owned small and medium-sized enterprises (SMEs). The largest company has about 430 employees and the many slaughterhouses (64 for pigs) are highly resistant to crisis and competition. This is because family and company are mutually dependent, which implies that new family capital is used in times of crisis and high competition.

Total gross meat production accounts for 1.7 billion tonnes and pigmeat represents 61.4% of the total (Table 5.1). With a self sufficiency ratio of 196% for pigmeat in 1994, the Belgian meat sector is a net exporter. Traditionally, the exporters are oriented to other member states of the EU, mostly Germany, France and Italy. Pigmeat accounts for 61.1% of total Belgian meat exports. Total apparent meat consumption reached a level of 99.2 kg per person in Belgium in 1994 and the share of pigmeat represents 49.9%.

Since Belgium is a net exporter of pigmeat, the sector contributes positively to the national trade balance. The balance of trade surplus of pigmeat accounted for 27.9 billion BF in 1993 and represented 7.6% of the total balance of trade surplus for the Belgian economy.

Table 5.1 Meat balances for pigmeat and total meat (1994) in 1000 tonnes carcass weight equivalent

Tonnes carcase weight	Pigmeat	Total meat	Share pigmeat (%)
Gross production	1021.2	1663.3	61.4
– live exports	53.0	123.4	
+ live imports	51.1	143.0	
Net production	1019.3	1682.9	60.6
+ beginning stock	0.0	0.5	
– final stock	0.0	0.0	
– meat exports	582.1	953.1	61.1
+ meat imports	83.4	313.4	
Human consumption	520.6	1043.7	
(kg/person)	49.5	99.2	49.9
Self sufficiency %	196.1	159.3	

Source: Landbouw Economisch Instituut, 1995.

5.2   Methodology

First, the methodology used to measure the competitiveness of the Belgian pigmeat sector is explained. Secondly, the different elements of Porter's diamond are discussed in relation to the Belgian pigmeat sector.

5.2.1   Measurement

In order to realize the objectives, the study starts with an analysis of the current situation within the sector. A financial analysis, based on published annual accounts, is made for the years 1991, 1992 and 1993. A distinction is made between slaughterhouses and cutting units. The financial results of the largest private companies are aggregated (Table 5.2). This sample represents about 30% of total private pig slaughterhouses in Belgium in 1994, which represent about 83% of all slaughterings in 1994.

The financial analysis develops a model using ratios, which discriminates between failing and succeeding companies. The model is based on multiple linear discriminant analysis (Viaene and Gellynck, 1995). The statistical construction of the model is not a topic for this paper, but its application to the meat sector provides interesting information for further identification of problems in the sector.

To measure competitiveness on external markets, the refined BCG matrix is used by measuring market attractiveness and market position (see Chapter

Table 5.2 Number and type of company included in the financial analysis of the pigmeat sector

Type of company	Number of companies
Slaughterhouses	10
Cutting units	10
Total	20

1) on the German, Italian and French markets. This is analysed from the point of view of both the importing and exporting countries for the period 1987–1993. Market attractiveness of the foreign market is determined by market size and market growth. Market size corresponds to the volume of each of the individual products imported into the foreign market in 1993. Market growth corresponds to the percentage change for each product in the imported volume during the period 1987–1993. In the same way, the position of Belgian exports is determined in the matrix.

Market position on the foreign market is determined by market share and market share growth. Market share corresponds to the share of each of the individual products in total imports of the products under consideration in this analysis in each foreign market in 1993. Market share growth corresponds to the percentage change in each product's market share during the period 1987–1993. In the same way, the position of Belgian exports is determined in the matrix.

The data are based on the combined nomenclature used in external trade statistics by Eurostat. The market for each country is defined as the total volume imported of the following categories of pigmeat, both fresh and frozen:

- carcasses
- hams
- shoulders
- loins
- bellies
- deboned meat.

5.2.2 Explanation

To explain and identify the causes of the current situation in the Belgian pigmeat sector, Porters' diamond is used. The information required to evaluate the competitive position of the pigmeat industry was obtained by contacting 19 slaughterhouses and/or cutting units. During an intensive interview with the general manager, all relevant aspects were discussed. The interviews were based on a topic list which considered the characteristics of the company, purchase of raw materials, the production process, the level of sales and future prospects.

5.3 Results

5.3.1 Measuring competitiveness

Financial analysis Generally, the financial position of slaughterhouses and cutting units was weak during the period 1991–1993 and no improvement appears in the annual accounts. The main results of the financial analysis for 1993 are summarized in Table 5.3.

Table 5.3 Main results of financial analysis for the pigmeat sector, 1993

Financial ratios	Slaughterhouses	Cutting units
Liquidity		
Current ratio	0.98	1.11
Stock turnover	36	34
Credit customers–suppliers (days)	+3	+7
Solvency		
Degree of debt (%)	80	77
Self-financing degree (%)	2.7	6.2
Net result/financial cost	0.70	1.14
Profitability		
Net sales margin (%)	0.9	0.5
Business assets turnover	4.7	4.8
Net profits/own funds (%)	−4.7	−0.4
Financial leverage	−1.0	0.6
Value added %		
Labour costs	58	69
Depreciation	26	22
Financial costs	15	9
Taxes	4	5
Added profits/losses	−3	−5
Total	100	100

Liquidity ratios measure a company's ability to pay its current liabilities as they mature and also the degree of efficiency in using resources. The current ratio is computed by dividing current assets by current liabilities.[1] The stock turnover measures the level of stocks in relation to total sales and is computed by dividing the value of sales by the value of stocks. Creditors' payment period minus debtors' collection period measures, in days, the difference between the average payment period of accounts receivable and the average collection period of accounts payable.

The liquidity of the companies analysed is not quite high enough in the case of slaughterhouses and just high enough for cutting units to fulfil their short term payment liabilities. Pressure on stock turnover could lead to liquidity problems. The negative net working capital[2] of pig slaughterhouses is particularly alarming. Both slaughterhouses and cutting units provide longer credit to customers than they receive from suppliers. It means that debts towards suppliers have to be financed by other resources than sales. These other resources consist mostly of bank loans, which are much more expensive than credit from suppliers.

There are two types of solvency ratio. The first measures the extent to which a company finances its activity with debt as opposed to equity. The degree of debt measures the percentage of total funds provided by debts, while the self-financing degree is the percentage of total funds provided by retained earnings. The second type of ratio measures the company's ability

1. Current assets = cash + accounts receivable + stock + prepaid expenses; current liabilities = accounts payable + bank loans payable + accrued taxes + currents long-term debt.
2. Net working capital = current assets - current liabilities.

to generate a level of income sufficient to meet its debt obligations (net result/financial cost).

Pigmeat firms are characterized by a high degree of debt. It means that the share of own funds in total assets is very low (varying from 20% to 23%). Moreover, the degree of self-financing is low and declined during 1991–1993. This situation means that banks will be more and more careful in lending money because own funds are not increasing. In the case of slaughterhouses, the net result is not high enough to cover financial costs. It means that companies are unable to pay for new debts.

Profitability ratios provide an overall evaluation of the performance of a company and its management. It concerns the measurement of the returns generated by the company from several different aspects:

- **net sales margin**: an assessment of the overall profitability of a business by comparing the net trading profit achieved relative to the level of sales
- **business assets turnover**: sales divided by business assets[3]
- **net profits/own funds**: earnings after taxes divided by own funds, or the ability of a company to remunerate its shareholders;
- **financial leverage**: the relation between profitability of own funds and the profitability of total assets before taxes. Assets financed by debts imply financial costs and if the profitability of investments is lower than financial costs, the financial leverage ratio is lower than 1 and vice versa.

Profitability is under pressure because of low net sales margins (<1%). If the companies' business assets turnover declines, for example because of new investment, profitability will continue its decline and liquidity problems could occur. This situation is made worse by the negative financial leverage in the case of slaughterhouses. It means that financial costs are higher than the profitability of investments. The ratio net profits/own funds is negative because of losses in 1993.

By analysing value added, it becomes clear that labour costs are high and rising. Gross value added per employee is low and varies between 50 000 and 55 000 ECU per year during the period 1991–1993.

Table 5.4 Evolution of Belgian exports of pigmeat by destination, 1987–1993 in tonnes and %

Destination	1987		1993	
	Tonnes	%	Tonnes	%
Germany	59 979	21.2	218 218	55.3
France	113 878	42.7	65 137	16.5
Italy	64 996	26.0	75 741	19.2
Total	238 853	89.9	359 096	91.0
Total exports	253 454	100	394 473	100

Source: Based on Eurostat.

3. Business assets = fixes assets + stocks + accounts receivable.

Table 5.5 Evolution of market shares in Germany, Italy and France, 1987–1993, in % of the total imported volume of pigmeat

Importing country	Germany		Italy		France	
Exporting country	1987	1993	1987	1993	1987	1993
Belgium	12	28	14	13	35	23
Netherlands	61	45	49	40	38	32
Denmark	19	18	13	16	20	32
Total (%)	92	91	76	69	93	87
Total market × 1000 T	458	714	460	572	316	269

Source: Based on Eurostat.

The weak financial position of the slaughterhouses and cutting units is confirmed by applying the discriminant model. It results in a low score, indicating the presence of companies in the sample for which the risks of bankruptcy in the near future is a reality.

Refined BCG matrix on export markets. Belgium is an important exporter of pigmeat. In 1993, total exports of pigmeat reached a level of 394 473 tonnes (Table 5.4). The EU accounted for 99.4% of total Belgian pigmeat exports in 1993. The main importers were Germany, Italy and France, with a combined share of 91.0% in total exports.

During the period 1987–1993, Belgian pigmeat exports increased by 55.6%. The destination of exports changed dramatically with sales to France falling in absolute as well as proportional terms, while German sales rose dramatically and it became the main importer of Belgian pigmeat.

Trends in the Belgian market share and that of its competitors on the German, Italian and French market are illustrated in Table 5.5. The Netherlands is the main competitor on all three markets, but its market share has decreased, especially on the large German market. On the Italian and French markets, Belgium holds third position.

During the period 1987–1993, both the German and Italian markets increased, by 55.9% and 24.3% respectively. The sharp increase in German imports is mostly related to a decline in domestic production. The increase on the Italian market results from growing consumption while production remained constant. French imports decreased by 14.9% during the same period. France became self-sufficient in pigmeat in the first half of 1994.

The refined BCG matrix for the German market is shown in Table 5.6. The way market attractiveness and market position are defined and calculated is illustrated in the appendix to this chapter. On the German market, carcasses, hams, shoulders and deboned meat have a high market attractiveness/position combination. Belgian supplies correspond perfectly with this trend for carcasses, hams and deboned meat. This is why the Belgian pigmeat sector is competitive for these products on the German market.

These three product types represent 82% of total German imports in 1993. The high level of imports of carcasses is unique in the EU. It is relat-

Table 5.6 Market attractiveness and market position of the Belgian pigmeat industry on the German market, 1987–1993

Market attractiveness		Market position Low	Medium	High
High	G	Loins Shoulders **	Carcasses Hams ***	Deboned meat ***
	B	**	Deboned meat Hams ***	Carcasses ***
Medium	G	Bellies *	**	***
	B	Shoulders *	Bellies **	***
Low	G	*	*	**
	B	*	Loins *	**

G, German demand; B, Belgian supply.
Market attractiveness/position: ***, high; **, medium; *, low.

ed to the new and large cutting units, recently built in Germany, and the lack of live pigs. For loins, shoulders and bellies the competitive position of the Belgian pigmeat sector does not correspond perfectly to German demand.

- for shoulders, the Belgian supply is low and the German demand is medium
- for loins, market growth is low for Belgian supply and high for German demand (a weak competitive position)
- for bellies, market share growth is high for Belgian supply and low for German demand (a stronger competitive position).

Competition on the German market is sharp and originates mostly from within Belgium. German companies appreciate the quality of the Belgian pig as well as the service and flexibility of Belgian companies. Service and flexibility refers to quick and accurate response to changes in requirements of German customers. These changes in requirements concern topics such as packaging, time and place of delivery, payment period. Flexibility is also related to the willingness of Belgian companies to work late and during weekends in order to prepare late orders placed by German customers.

In order to create a sustainable relationship between supplier and customer, German buyers aim to set contracts where both price and quantity are specified. However, this causes problems for Belgian companies because of the lack of price transparency at the supply side, especially for live pigs. There are strong and frequent price variations.

The Dutch and Danish pigmeat sectors are characterized by high concentration. Large companies can more readily fulfil the needs of the largest buyers such as supermarket chains and large meat-product companies in Germany. Belgian exporters operate much more in niche markets, where price competition is not as high as on mass-markets.

In 1995, France started exporting to the German market. This means that France will become an additional competitor on the German market in the coming years.

Italian pig farming is concentrated in the north; the Po valley area accounts for more than 70% of total Italian pig production. Italian pig farming is mostly oriented towards production of the famous Parma ham. However, Italian pigmeat production only reached a self-sufficiency level of 66% in 1993 (MLC, 1996).

Table 5.7 shows market attractiveness/position combinations on the Italian market. The Italian market is interesting for hams and deboned meat. The Belgian competitive position corresponds to the Italian demand for hams, but not for deboned meat. In relation to deboned meat, market growth in Belgian exports is low, while Italian demand increased during the period 1987–1993.

On the Italian market for carcasses, Belgian exporters have a highly competitive position. Despite the fact that the Italian demand results in a low

Table 5.7 Market attractiveness and market position of the Belgian pigmeat industry on the Italian market, 1987–1993.

Market attractiveness		Market position		
		Low	Medium	High
High	I	**	***	Hams ***
	B	**	***	***
Medium	I	*	**	Deboned meat ***
	B	*	**	Carcasses Hams ***
Low	I	Loins Bellies Carcasses *	Shoulders *	**
	B	Shoulders Bellies *	Deboned meat Loins *	**

I, Italian demand; B, Belgian supply.
Market attractiveness/position: ***, high; **, medium; *, low.

market attractiveness/position combination, the Belgian pigmeat industry obtains a high market attractiveness /position combination. It relates to the medium level of market growth of Belgian exports, while the market growth of Italian demand is low. Belgian slaughterhouses export relatively more carcasses rather than added-value deboned meat.

Considering the other products (bellies, loins and shoulders), the Belgian market attractiveness position corresponds with Italian demand, but for all three products the market attractiveness/position combination is low.

During the period 1990–1994, French self-sufficiency degree for pigmeat increased by 14% and France became self-sufficient in 1994. Pig farming in France is highly concentrated in Brittany, which represents 53% of the total French pig population (MLC, 1996).

The Belgian competitive position on the French market weakened during the period 1987–1993, particularly for hams (see Table 5.8). In general this is related to low market size and market growth of Belgian exports, while French demand is characterized by a high market size and medium market growth. Belgian hams lost their strong position to Danish companies. The heavy weight of Belgian hams is a disadvantage (Viaene and Gellynck, 1989).

For carcasses, loins and deboned meat, the Belgian pigmeat sector is competitive as the market attractiveness/position combination of Belgian exports corresponds with the market attractiveness/position combination of French imports. A weak competitive position is noticed for shoulders. It results from

Table 5.8 Market attractiveness and market position of the Belgian pigmeat industry on the French market, 1987–1993

Market attractiveness		Low	Market position Medium	High
High	F	**	***	Hams ***
	B	**	***	Bellies ***
Medium	F	*	Shoulders **	***
	B	*	Deboned meat **	***
Low	F	Loins *	Carcasses *	Deboned meat Bellies **
	B	Carcasses Loins *	Shoulders *	Hams **

F, French demand; B, Belgian supply.
Market attractiveness/position: ***, high; **, medium; *, low.

high market growth of French imports, while Belgian exports obtain only a medium market growth.

For Belgian exporters, meat quality and service is the basis of competitiveness whereas with the Netherlands and Denmark, price dominates (Viaene and Gellynck, 1988). However, on all three markets sales prices are under pressure because of overcapacity in the Netherlands, Germany and France.

5.3.2 Explaining competitiveness

Factor conditions. Live pigs are the raw material for slaughterhouses and relations with suppliers of live pigs is based on a specific purchase pattern, trust and tradition. However, it is surprising that almost every slaughterhouse has its own purchasing pattern, and these vary widely:

- direct supply by the producer
- payment based on live- or deadweight
- with or without a premium for quality
- with or without the use of a trade agent as intermediary
- with cash payment or up to 5–6 weeks' credit.

These varying criteria and their many possible combinations result in a lack of price transparency and difficulty in price determination. Based on the interviews, it became clear that the most profitable slaughterhouses are those with a standard purchasing pattern focusing on a specific type of pig.

Cutting units purchase live pigs, carcasses or parts. The share of live pigs in total purchases depends on the degree of specialization, price, physical integration and special customer requirements. A high degree of specialization means that companies work with a limited number of products, and results in few purchases of live pigs. Related to price, it is often cost efficient to purchase the desired parts (e.g. hams) rather than live pigs where remaining unwanted parts must be disposed of. Physical integration relates to the fact that some cutting units are linked to slaughterhouses. It often means that contracts exist to slaughter a minimum of live pigs per year. It also happens that special customer requirements force a slaughterhouse or cutting unit to purchase specific parts.

Three types of problem occur in relation to purchases of raw materials:

- lack of price transparency
- uncertainty about the continuity of supply
- quality of the purchased raw material.

The lack of price transparency for live pigs results in a high level of competition among slaughterhouses. Attempts at co-operation and agreements between slaughterhouses and cutting units have not thus far been successful. Due to overcapacity, it is extremely difficult to obtain a competitive advantage through a sharper purchasing policy. Despite these difficulties, those interviewed consider the present method of price determination to be balanced and acceptable.

Uncertainty about the continuity of supply is related to the overcapacity problem and to the pressure on prices for live pigs. Under these conditions, pig suppliers are not always reliable. Slaughterhouses and cutting units thus focus on a win–win situation, by guaranteeing purchase.

Raw material quality problems relate to pale, soft and exudative (PSE) meat and to residues. During the last 10 years, a lot of work has been done to avoid the PSE problem which has been substantially ameliorated. More recently the control of antibiotic residues has been targeted and the proportion of positive tests has dropped significantly.

The most important component of production cost is the live animals. As already indicated, overcapacity in slaughterhouses has resulted in high demand and prices for live pigs. The second element consists of labour costs, which are higher in Belgium than in its European competitors. High taxes and levies for social security are the basis of this problem. Three of the interviewed companies are looking for a location outside Belgium where, besides labour costs, environmental legislation and the interference of public authorities are less severe and less expensive. In order to control labour costs and to maintain productivity, the interviewed companies search for flexibility in the number of persons employed by applying 'technical unemployment' and compulsory leave.

Because suppliers are usually paid within a week, whereas customers take 30–60 days to pay, financial costs are important. It means that purchases of raw materials have to be financed by other resources than sales. Slaughterhouses are capital intensive and this implies high fixed costs which, combined with overcapacity, impose a heavy burden on company balance sheets.

These combined problems mean that some slaughterhouses do not know the production cost of the products they sell. Companies work in an intuitive way and undercut the market. The best performers compute a cost both per product and per customer on a weekly basis, thus enabling accurate production and sales management.

Some slaughterhouses score low on productivity of capital, due to overcapacity. Because of the variable activity of slaughterhouses, labour productivity could be improved by increasing flexibility. Productivity of labour is crucial and, because of overcapacity, labour productivity is under pressure.

The products of slaughterhouses and cutting units are distinguished in increasing order of transformation:

- carcasses
- first cut: bellies, hams, loins, shoulders
- second cut: deboned and defatted
- third cut: portioned
- fourth cut: preparations such as cooked dishes.

Each of these product types require specialized technology. Technology in Belgian slaughterhouses and cutting units is well developed at the levels of carcasses and first and second cuts, but generally poor at the levels of third and fourth cuts. However, it is noticeable that companies which specialize in a certain activity, independent of the level of transformation, manage to be profitable.

At the end of the 1980s investments were high in the Belgian meat industry, partly supported by the European Orientation and Guarantee Fund for Agriculture (FEOGA). These investments were not limited to the improvement of production techniques, but also included the extension of capacity. Today, these investments result in an estimated overcapacity of 30% in slaughterhouses and cutting units in the pigmeat sector. Companies which intend to invest during the period 1995–2000 foresee the need to extend cutting capacity for the second and third cuts. However, some companies intend to invest in 'software' rather than in 'hardware'. These companies are convinced that during the last 10 years too much has been invested in buildings and trucks, but not enough in hygiene, quality and management. The objective is to improve profitability and quality and especially to produce what the market requires.

Analysing the link between investment and profitability, three elements are clear:

- companies that invested before 1990 realize higher profitability than companies which invested after 1990
- the largest and smallest companies are relatively more profitable than the group in the middle
- companies working with trained and experienced managers score better than those run by managers close to retirement or by managers who are too young and inexperienced to run a company in times of crisis.

The sector requires management innovation given the changing working conditions. During the 1970s and 1980s, margins were high and the cost of errors could be relatively easily absorbed. Today, the sector faces low margins and errors are catastrophic. It means that accurate control and feedback is a necessity.

Demand conditions.   The share of food expenditures in total expenditures is low and falling in Belgium. During the period 1980–1992, this share declined by 17.4% to 14.7%.

Demographic changes are an important factor for meat consumption. As elsewhere in western Europe, Belgium has a stagnant, ageing population. This does, however, imply that the category of people consuming most meat has increased. In 1990, people over 65 years old consumed about 45.7 kg per capita, while people less than 36 years old consumed about 26.8 kg per head (Table 5.9). However, the longer run outlook is not bright because younger people not only consume less meat in general but in particular consume a significantly lower level of pigmeat (though their share of pigmeat in total meat consumption is higher than for older people).

Table 5.9 Household consumption of meat in Belgium according to age (1990) in kg per person

Type of meat	<36 years	36–50 years	51–65 years	>65 years	Total
Beef	8.0	10.4	14.8	16.4	12.4
Pigmeat	10.2	12.3	14.4	14.6	13.1
Poultry	8.6	10.7	15.1	14.7	12.4
Total	26.8	33.4	44.3	45.7	37.9

Source: LEI panel, 1990.

Continuous negative publicity about meat (news in the mass media about scandals relating to hormones, antibiotics, residues or animal welfare) makes it very hard to project a positive image for meat.

Consumer needs are dominated by two important factors: increasing demand for

- variety and value added products
- safe meat, which in Belgium refers especially to the absence of antibiotic and hormone residues.

Within the framework of these changing demand patterns, it is difficult for the Belgian meat sector to develop a brand identity at the consumer level. The development of a brand is capital intensive (research and advertising) and risky, and margins on the one hand and the structure of the Belgian pigmeat sector on the other hand (SMEs) do not permit these investments.

At the industrial sector level, Belgian slaughterhouses have a good reputation for:

- high quality products combined with good service
- flexibility related to the requirements of the customer.

Related and supporting industries. The following types of customer can be distinguished:

- retail sector
- processed meat companies.

Under pressure from the increasing bargaining power of the retail sector, slaughterhouses are forced to produce more value added products, which are not always fully compensated by a higher price. Each customer also requires specific service, related to the product and packaging as well as to time of delivery and payment. Prices are determined after delivery of the product. Co-operation with suppliers is essential for the retail sector in order to avoid dependence on a single supplier. Retailers stimulate several companies to invest in further finishing of products. However, the retail sector is considered to be a fair partner and correct payer by the few companies able to fulfil their requirements.

Developments in the processed meat sector are similar to those in the retail sector. However, processed meat companies switch more often between suppliers. Processed meat companies also make price agreements among themselves. Payment is somewhat difficult because of problems in the processed meat sector, related to declining consumption and their own over-capacity.

Firm strategy, structure and rivalry.   During the period 1988–1994, the number of pigs slaughtered in Belgium increased by 15.8% to reach 10 768 954 in 1994. During the same period, the number of slaughterhouses declined by 25% (Table 5.10). Obviously more animals are being slaughtered by fewer companies.

There are both private and public slaughterhouses in Belgium. The public slaughterhouses operate as service companies, mostly for cutting units. The largest slaughterhouses are in private hands and represent 83% of total slaughterings. Despite increasing concentration, Belgian slaughterhouses remain small compared to the Dutch, French and Danish.

Since products from different companies are good substitutes (product differentiation is difficult to realize) most companies use price as the only competitive weapon. This results in a high level of price competition. However, price is not the only element taken into consideration by the customer who also looks for service.

Co-operation in the sector is very low. Attempts in the past have failed and harmed confidence. However, current problems in the sector have created a new willingness to co-operate at the level of price determination, export to non-EU markets and quality control (Hazard Analysis Critical Control Point and the International Standards Organisation, for example).

Potential competition of new entrants is high because products are homogeneous and entry barriers are low. If one firm specializes, for example in third cut, it can be easily copied by colleagues with relatively low investment. Moreover, all competitors work with the same basic product and brand identity does not exist. The threat of substitutes is also a reality. Both poultry and fish are increasing their market shares at the expense of beef and pigmeat.

Voluntary exit from the sector is rare: a strong financial as well as intense emotional link exists because companies are family owned. This is reinforced

Table 5.10 Structure of the Belgian pig slaughterhouses with more than 10 000 slaughterings per year, 1988–1994

Slaughterings per year	Slaughterhouses				Slaughterings			
	Number		%		× 1000		%	
	1988	1994	1988	1994	1988	1994	1988	1994
10 000 – 100 000	29	15	45.3	31.2	1080	545	11.6	5.1
100 000 – 300 000	26	20	40.6	41.7	4337	3753	46.6	34.8
> 300 000	9	13	14.1	27.1	3883	6470	41.8	60.1
Total	64	48	100	100	9301	10 769	100	100

Source: Ministry of Agriculture and IVK.

by the fact that slaughterhouses and cutting units have no alternative uses. Moreover, the very expensive regulations in Belgium for dismissal of employees make closing down almost impossible.

Just as in Belgium, the meat sectors in Germany, the Netherlands and France are confronted by overcapacity, which has also partly been supported by EU funds. This situation creates an additional supply and price pressure.

The EU as a whole is self-sufficient in pigmeat and exports to third countries are vital for the price level on the internal market. As, for example, Danish exports of pigmeat to Japan decrease, the decline in exports is available for sale on the European market and results in price pressure. In the future, it will be important for the Belgian pigmeat sector to look for potential markets outside the EU. Since more than 60% of Belgian exports are realized in one market (Germany), Belgium occupies a very vulnerable position in the international pigmeat market.

Profitability of companies is not linked to the percentage of sales realized on the domestic or external markets. The most profitable companies are characterized by a stable and clear sales pattern, focused on specific products to fulfil specific requirements, whatever the market.

Government. The impact of government involvement is felt in two different ways:

- effects of the single European market (SEM) and Common Agricultural Policy (CAP);
- effects of the General Agreement on Tariffs and Trade (GATT). On the domestic market, the creation of the SEM has had two major consequences:
- slaughterhouses were obliged to invest to meet European export hygiene standards
- French and German retail chains increased competition and pressure on pigmeat prices at the level of slaughterhouses and cutting units.

Increased competition from other member states was not experienced.

Slaughterhouses with activity limited to the domestic market were not obliged to invest, and could continue to work until 1994. Some of these companies are still active and destabilize the market because of their lower production costs. It is up to the Commission to insist that national governments in Belgium and other member states apply EU regulations.

In relation to the CAP, export restitutions do not stimulate the export of high value added products such as deboned meat. As Belgium specializes in deboned meat, this increases competition on value added products. Given the changing patterns of demand, restitutions should be changed to stimulate exports of value added rather than mass products. This suggestion, emanating from the slaughterhouses, refers to the fact that the Commission decided to eliminate restitutions for deboned meat and maintain them for carcasses and other low value added products.

All of the exporting slaughterhouses interviewed stated that the creation of the SEM had little impact on their exports. Administration and transportation delays were reduced because of less frontier bureaucracy (Nichols *et al.*, 1995; Viaene and Gallet, 1989). It also appears that harmonization rather than mutual recognition had some effect, while national preferences and prejudices remain: for example, the negative publicity in the German press about the way meat is controlled in Belgian and Dutch slaughterhouses.

The GATT agreement was first implemented in July 1995, with changes being gradually applied over 6 years (10 years for developing countries). According to the slaughterhouses, increased competition on third markets from the US will decrease the market share of the EU. It means that the surplus will be offered for sale on the SEM, which will increase competition and put additional pressure on sales prices.

The main elements of the GATT agreement are:

- reduction in export restitutions by 36% and subsidized export volumes by 21%, up to the year 2000
- reduction of domestic support by 20%
- minimum import access must account for at least 3% of 1986–1988 internal consumption from the beginning of the transition and must increase to 5% at the end of it
- tarification of border protection: all border measures must be transformed into tariff equivalents, and these equivalents must be reduced by 36% over a 6 year period.

The implementation of the GATT agreement in the pigmeat sector has considerable consequences on both imports and exports. For imports, the variable levies are replaced by tariff equivalents based on the average levies in 1986–1988. The tariffs must be reduced in six equal annual steps by a total of 36% by the year 2000 (Table 5.11).

From 1995 to 1997, the tariff equivalents are quite high and border protection will be at the same level. Special safeguard mechanisms are authorized under the GATT agreement if:

- import volume increases strongly
- import prices fall below reference price levels.

Table 5.11 Tariff equivalents in ECU per tonne

Pigmeat	1 July 1995	2000
Carcasses	788	504
Hams	1 142	731
Shoulders	883	565

Source: EU Commission and authors' calculations.

For pigmeat, minimum market access increases by 12 500 tonnes per year from 63 500 tonnes in 1995 to 75 600 tonnes in 2000. Exportation of pigmeat with restitutions from the EU-12 decreases from 491 000 tonnes in 1995 to 402 000 tonnes in 2000. Also, the commitment level regarding the budget declines from 183.4 million ECU in 1995 to 117.4 million ECU in 2000.

The GATT agreement will result in a sharp decrease in pigmeat exports receiving restitutions (Figure 5.1). This decline in exports will lead to a repositioning of exports between EU member states, having a negative impact on Belgium, since Belgian exports are concentrated on a few member states.

Dynamics. A strong interaction of the positive determinants of their competitive position should result in competitive advantage. The strength of this interaction depends on two factors (Douglas and Craig, 1995):

- geographical concentration, which accelerates the diffusion of innovation, development of specialized resources and of supporting industries
- the development of clusters, which facilitate the creation of advanced factors such as specialized technology and skills, while the success of downstream industries stimulates growth in supplier industries and vice versa.

However, in the case of the Belgian pigmeat sector the interaction between the different components limits the creation of competitive advantage rather than stimulates it. The pigmeat sector, including feed companies, farmers, slaughterhouses, cutting units and processed meat plants, is geographically concentrated in the Flemish part of Belgium, but this does not

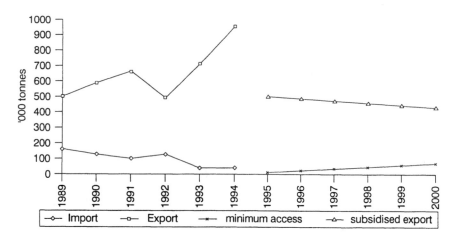

Figure 5.1 Expected evolution of imports and exports of pigmeat EU-12 1989–2000 (tonnes) ◇, import; □ export; ×, minimum access; △, subsidized export. Based on Eurostat and author's calculations.

accelerate innovation at the level of slaughterhouses and cutting units. A slight pull effect from processed meat companies and the retail sectors stimulates some companies to innovate at the management level. However, the majority of companies do not change management and are not able to manage change. This results in unworkable (price) competition and market power in the hands of suppliers and customers.

The development of advanced factors such as specialized technologies and skills is hampered by the structure of the industry. This structure implies a lack of capital to develop the advanced factors and ultimately competitive advantage.

5.4 Conclusions

The poor financial situation of Belgian pig slaughterhouses and cutting units is caused by a complex set of factors, which are determined by both local and international components.

On the domestic market, problems are situated at the sectoral and governmental level as well as at the level of supply and demand. The sector is characterized by SMEs and a low level of concentration in combination with overcapacity, which results in price pressure. Investments to meet European export hygiene standards were stimulated by EU subsidies.

Lack of transparency in live animal price determination and a uniform end product result in intensive competition and pressure on margins. These changing working conditions characterized by small margins make calculation of production costs essential.

On the demand side, both the evolution of population and consumption are stagnant or even decreasing. Moreover, meat suffers from a negative image. It means that slaughterhouses and cutting units work on a stagnating or even declining domestic market.

In export markets, Belgium has maintained its position due to product quality and service. However, price competition has increased because of overcapacity of slaughterhouses and cutting units in neighbouring countries (the Netherlands, Germany and France). From the results of the refined BCG matrix, it becomes clear that Belgium is not always able to maintain its competitive position on the main export markets. Especially on the French market the competitive position of the Belgian pigmeat sector has weakened.

The SEM seems to have had little effect, while more impact is expected from the GATT. Export restitutions will decline and increased competition from the USA is expected on third markets, especially South-East Asia. In this way an additional volume of meat presently sold in this area will be lost and offered for sale in Europe, which will put pressure on prices.

At the company level, it is necessary that management adjust to face the future. Costs have to be reduced and controlled to increase productivity of both labour and capital. Before investing in further transformation, slaughterhouses and cutting units should first acquire a leading position in their current activity. Investment in people, rather than in capacity or further transformation, is a priority.

On the demand side, 'healthy meat' is required by the consumer. This means that the whole production chain has to work together to provide guarantees to the consumer about meat quality and production methods.

The Commission should continue to insist that national governments apply Community Regulations related to export standards, and no exceptions should be admitted. Companies which have not modernized should stop their activities.

The problem of overcapacity is not limited to the domestic market. It should be solved at the European level, in co-operation with the Commission and the sector. By oversubsidizing investment, the Commission is partly responsible for the present situation.

Since natural elimination of companies in difficulty would be slow and also harm profitable companies, it is necessary to work out a restructuring plan for the sector as quickly as possible. The aim of this plan should be to buy out capacity.

Related to the implementation of the GATT, the Commission should take care to stimulate exports of value added products rather than mass products. In the case of the meat business, this refers especially to deboned meat, which must currently be sold on third markets without restitutions.

As exports to third markets remain vital for the pigmeat industry, Belgian companies should work together to finance market research and organize common exports. Special attention should be given to the developments in eastern Europe.

5.4.1 Benefits and disadvantages of a Porter analysis of the sector

By using Porters' diamond to determine the competitive position of the Belgian pigmeat sector, it has become clear that this approach has both strengths and weaknesses. A major strength of Porters' diamond is that it provides a practical checklist to analyse a sector. The emphasis on mutual relationships between each component of the diamond is especially useful.

One major weakness of the Porter model is related to the fact that it does not provide a way to measure competitiveness of a sector objectively. Its value is limited to that of an analytical tool. Interpretation by the researcher becomes extremely important and may even become subjective. Since the model is rather complete and considers all the factors influencing competitiveness, focusing on one specific factor gives the possibility of classifying a sector as either competitive or uncompetitive.

A second major weakness is that within the Porter model one component can be so dominant that it counterbalances other components and results in an uncompetitive sector, even if the other points of the diamond suggest that the competitive position is strong (or vice versa). This is particularly the case when a competitive sector, defined as 'one that possesses the sustained ability to profitable gain and maintain market share in foreign markets' (Agriculture Canada, 1991) is characterized by a high level of government intervention, which dominates the other components. In the European food industry the meat sector is a clear example of this.

Appendix: Market attractiveness and market position for pigmeat on the German, Italian and French markets

Table 5A.1 Evolution of market size in Germany, Italy and France, 1987–1993, in total imported tonnes of pigmeat

Products	Germany		Italy		France		Total	
	1987	1993	1987	1993	1987	1993	1987	1993
Carcasses	249 405	331 709	84 789	85 637	133 889	49 539	468 083	466 885
Hams	27 367	52 544	256 256	353 633	86 244	105 218	369 867	511 395
Shoulders	27 880	56 038	21 453	24 027	8 566	16 618	57 899	96 683
Loins	34 539	46 282	5 694	5 808	4 945	2 077	45 178	54 167
Bellies	20 644	28 489	22 092	23 418	21 539	26 924	64 275	78 831
Deboned meat	97 472	198 843	68 277	79 202	61 999	68 199	227 748	346 244
Total	457 307	713 905	458 561	571 725	317 182	268 575	233 050	1 554 205
Average	76 218	118 984	76 427	95 287	52 864	44 762	205 508	259 034
Average (three countries)	–	–	–	–	–	–	68 503	86 345

Source: Based on Eurostat data.

Table 5A.2 Evaluation of market size per pigmeat product in Germany, Italy and France, 1993 in imported tonnes. High market size (+) ≥ 86 345 tonnes, medium market size (±) < 86 345 and > 43 172 tonnes, low market size ≤ 43 172 tonnes

Products	Germany	Italy	France
Carcasses	+	±	±
Hams	±	+	+
Shoulders	±	–	–
Loins	±	–	–
Bellies	–	–	–
Deboned meat	+	±	±

Source: Based on Eurostat data.

Table 5A.3 Market growth on the German, Italian and French pigmeat markets, 1987–1993, in percentage imported tonnes. High market growth (+) ≥ 28%, medium market growth (±) < 28% and > 14%, low market growth ≤ 14%

Products	Germany		Italy		France	
	Market growth (%)	Evaluation	Market growth (%)	Evaluation	Market growth (%)	Evaluation
Carcasses	+33	+	+1	–	–63	–
Hams	+92	+	+38	+	+22	±
Shoulders	+101	+	+12	–	+94	+
Loins	+34	+	+2	–	–58	–
Bellies	+38	+	+6	–	+25	±
Deboned meat	+104	+	–16	±	+10	–
Average	+67		+12		+5	
Average market growth per product (three countries)						+28

Source: Based on Eurostat data.

Table 5A.4 Evolution of market share in Germany, Italy and France, 1987–1993, in % imported tonnes per product

Products	Germany		Italy		France	
	1987	1993	1987	1993	1987	1993
Carcasses	50.5	46.5	15.5	15.0	24.3	18.4
Hams	7.2	7.4	58.4	61.8	35.0	39.2
Shoulders	7.7	7.8	4.2	4.2	6.0	6.2
Loins	6.5	6.5	1.0	1.0	0.8	0.8
Bellies	4.0	4.0	4.1	4.1	9.7	10.0
Deboned meat	26.1	27.8	13.7	13.8	23.9	25.4
Total	102	100	96.9	99.9	99.7	100

Source: Based on Eurostat data.

Table 5A.5 Evaluation of market share in Germany, Italy and France, 1987–1993, in % imported tonnes per product. High market share (+) ≥ 16.6%, medium market share (±) < 16% and > 8.3%, low market share ≤ 8.3%

Products	Germany	Italy	France
Carcasses	+	±	+
Hams	–	+	+
Shoulders	–	–	–
Loins	–	–	–
Bellies	–	–	±
Deboned meat	+	±	+

Source: Based on Eurostat data.

170 COMPETITIVENESS IN THE FOOD INDUSTRY

Table 5A.6 Market share growth on the German, Italian and French pigmeat markets, 1987–1993, in percentage imported tonnes. High market growth (+) ≥ 0, low market growth ≤ 0

Products	Germany Market share growth (%)	Evaluation	Italy Market share growth (%)	Evaluation	France Market share growth (%)	Evaluation
Carcasses	–8.1	–	–3.5	–	–24.0	–
Hams	+1.4	+	+5.9	+	+12.0	+
Shoulders	–1.8	–	+0.4	+	+3.0	+
Loins	–1.1	–	–0.2	–	–0.8	–
Bellies	–0.5	–	–0.7	–	+3.0	+
Deboned meat	+6.5	+	+0.9	+	+6.0	+
Average	–0.6		+0.5		–0.1	
Average market growth per product (three countries)						+0.12

Source: Based on Eurostat data.

Table 5A.7 Market attractiveness for pigmeat products on the German market based on imported volume (demand)

Market growth

	Bellies	Hams Shoulders Loins	Carcasses Deboned meat
High	**	***	***
Medium	*	**	
Low	*	*	**
	Low	Medium	High

Market size

Market attractiveness: ***, high; **, medium; *, low.

Table 5A.8 Market position for pigmeat products on the German market based on imported volume (demand)

Market share growth

High	Hams **	***	Deboned meat ***
Medium	*	**	***
Low	Loins Bellies Shoulders *	*	Carcasses **
	Low	Medium	High

Market share

Table 5A.9 Market attractiveness for pigmeat products on the Italian market based on imported volume (demand)

Market growth

High	**	***	Hams ***
Medium	*	Deboned Meat **	***
Low	Shoulders Loins Bellies *	Carcasses *	**
	Low	Medium	High

Market size

Table 5A.10 Market position for pigmeat products on the Italian market based on imported volume (demand)

Market share growth

High	Shoulders **	Deboned meat ***	Hams ***
Medium	*	**	***
Low	Loins Bellies *	Carcasses *	**
	Low	Medium	High

Market share

Table 5A.11 Market attractiveness for pigmeat products on the French market based on imported volume (demand)

Market growth

High	Shoulders **	***	***
Medium	Bellies *	**	Hams ***
Low	Loins *	Carcasses Deboned meat *	**
	Low	Medium	High

Market size

Table 5A.12 Market position for pigmeat products on the French market based on imported volume (demand)

Market share growth

High	Shoulders **	Bellies ***	Hams Deboned meat ***
Medium	 *	 **	 ***
Low	Loins *	 *	Carcasses **
	Low	Medium	High

Market share

Table 5A.13 Evolution of market size in Germany, Italy and France, 1987–1993, in total exported tonnes of pigmeat from Belgium

Products	Germany		Italy		France		Total	
	1987	1993	1987	1993	1987	1993	1987	1993
Carcasses	12 491	109 820	6 001	10 430	55 647	10 598	74 139	130 848
Hams	1 096	10 194	31 916	41 780	8 783	7 573	41 795	59 547
Shoulders	2 222	7 258	2 385	841	3 119	5 299	7 726	13 398
Loins	11 376	14 103	643	837	4 325	2 044	16 344	16 984
Bellies	2 283	8 397	5 909	4 230	4 416	11 706	12 608	24 333
Deboned meat	30 511	68 446	18 142	17 623	37 588	27 917	86 241	113 986
Total	59 979	218 218	64 996	75 741	113 878	65 137	238 853	359 096
Average	9 997	36 370	10 833	12 624	18 980	10 856	39 809	59 849
Average (three countries)	–	–	–	–	–	–	13 270	19 950

Source: Based on Eurostat data.

Table 5A.14 Evaluation of market size in Germany, Italy and France, 1993, in exported tonnes from Belgium. High market size (+) ≥ 19 950 tonnes, medium market size (±) < 19 950 and > 9975 tonnes, low market size ≤ 9975 tonnes

Products	Germany	Italy	France
Carcasses	+	±	±
Hams	±	+	–
Shoulders	–	–	–
Loins	±	–	–
Bellies	■	■	+
Deboned meat	+	±	+

Source: Based on Eurostat data.

Table 5A.15 Market growth on the German, Italian and French pigmeat markets, 1987–1993, in percentage exported tonnes from Belgium. High market growth (+) ≥ 131%, medium market growth (±) < 131%, low market growth ≤ 65%

Products	Germany		Italy		France	
	Market growth (%)	Evaluation	Market growth (%)	Evaluation	Market growth (%)	Evaluation
Carcasses	+779	+	+74	±	−81	−
Hams	+830	+	+31	−	−14	−
Shoulders	+226	+	−65	−	+70	±
Loins	+24	−	+30	−	−53	−
Bellies	+268	+	−28	−	+165	+
Deboned meat	+124	±	−3	−	−43	−
Average	+375		+6		+7	
Average market growth per product (three countries)						+131

Source: Based on Eurostat data.

Table 5A.16 Evolution of market share in Germany, Italy and France, 1987–1993, in percentage of exported tonnes per product from Belgium

Products	Germany		Italy		France	
	1987	1993	1987	1993	1987	1993
Carcasses	20.8	50.3	9.2	13.8	48.9	16.3
Hams	1.8	4.7	49.1	55.2	7.7	11.6
Shoulders	3.7	3.3	3.7	1.1	2.7	8.1
Loins	19.0	6.5	1.0	1.1	3.8	3.1
Bellies	3.8	3.8	9.1	5.6	3.9	18.0
Deboned meat	50.9	31.4	27.9	23.3	33.0	42.9
Total	100	100	100	100	100	100

Source: Based on Eurostat data.

Table 5A.17 Evaluation of market share in Germany, Italy and France, 1993, in % of exported tonnes per product from Belgium. High market size (+) ≥ 16.6%, medium market share (±) < 16.6% and > 8.3%, low market share ≤ 8.3%

Products	Germany	Italy	France
Carcasses	+	±	±
Hams	−	+	±
Shoulders	−	−	−
Loins	+	−	−
Bellies	−	−	+
Deboned meat	+	+	+

Source: Based on Eurostat data.

Table 5A.18 Market share growth on the German, Italian and French pigmeat markets, 1987–1993, in percentage exported tonnes from Belgium. High market growth (+) ≥ 0%, low market growth ≤ 0%

Products	Germany		Italy		France	
	Market share growth (%)	Evaluation	Market share growth (%)	Evaluation	Market share growth (%)	Evaluation
Carcasses	+141.6	+	+49.1	+	−66.7	−
Hams	+155.6	+	+12.3	+	+50.7	+
Shoulders	−10.2	−	−69.7	−	+197.0	+
Loins	−65.9	−	+11.7	+	−17.4	−
Bellies	+1.1	+	−38.6	−	+363.4	+
Deboned meat	−38.3	−	−16.6	−	−29.8	−

Source: Based on Eurostat data.

Table 5A.19 Market attractiveness for pigmeat products on the German market, based on exported volume from Belgium (supply)

Market growth

High	Shoulders Bellies **	Hams ***	Carcasses ***
Medium	*	**	Deboned meat ***
Low	*	Loins *	**
	Low	Medium	High

Market size

Market attractiveness: ***, high; **, medium; *, low.

Table 5A.20 Market position for pigmeat products on the German market, based on exported volume from Belgium (supply)

Market share growth

	Low	Medium	High
High	Hams Bellies **	***	Carcasses ***
Medium	*	**	***
Low	Shoulders *	*	Loins Deboned meat **
	Low	Medium	High

Market share

Table 5A.21 Market attractiveness for pigmeat products on the Italian market, based on exported volume from Belgium (supply)

Market growth

	Low	Medium	High
High	**	***	***
Medium	*	Carcasses **	***
Low	Shoulders Loins Bellies *	Deboned meat *	Hams **
	Low	Medium	High

Market size

Table 5A.22 Market position for pigmeat products on the Italian market, based on exported volume from Belgium (supply)

Market share growth

High	Loins **	Carcasses ***	Hams ***
Medium	*	**	***
Low	Shoulders Bellies *	*	Deboned meat **
	Low	Medium	High

Market share

Table 5A.23 Market attractiveness for pigmeat products on the French market, based on exported volume (supply)

Market growth

High	**	Bellies ***	***
Medium	Shoulders *	**	***
Low	Hams Loins *	Carcasses *	Deboned meat **
	Low	Medium	High

Market size

Table 5A.24 Market position for pigmeat products on the French market, based on exported volume from Belgium (supply)

Market share growth

High	Shoulders **	Hams ***	Bellies ***
Medium	*	**	***
Low	Loins *	Carcasses *	Deboned meat ***
	Low	Medium	High

Market share

References

Agriculture Canada (1991) Task Force on Competitiveness in the Agri-Food Industry: *Growing Together: Report to Ministers of Agriculture*. Agriculture Canada, Ottawa.

Douglas, S.P. and Craig, C.S. (1995) *Global Marketing Strategy*, McGraw-Hill, Singapore.

Economist Intelligence Unit (1995) Is the single market working?, small and medium-sized enterprises give their verdict. *Business Europe*, 1 May.

LEI (1995) *Landbouwstatistisch Jaarboek 1994*, Landbouw Economische Instituut Statistieken, Ministerie van Middenstand en Landbouw, Brussels.

MLC (1996) *European Handbook: EEC and International Statistics*, Vol. II. Meat and Livestock Commission, Milton Keynes, UK.

Nichols, J., Sargent, M., Viaene, J., De Craene, A. and de Noronha Vaz, M.T. (1995) The internationalisation of small and medium sized food and drink firms, in *Europe in Progress: Model and Facts* (ed. S. Urban), Gabler Verlag, Wiesbaden.

Porter, M.E. (1990) *The Competitive Advantage of Nations*. Harvard University Press, Cambridge, MA.

Rabobank (1993) *Competitiveness in the Pig Industry*. Rabobank Nederland.

Viaene, J. and De Craene, A. (1988) Improvement of the Belgian export position for pigs and pigmeat. Division of Agro-Marketing, University of Ghent.

Viaene, J. and Gallet, G. (1989) Impact of 1992 on the food industry. National State Secretary for Europe 1992. Brussels, May.

Viaene, J. and Gellynck, X. (1989) Distribution structure and image of Belgian meat in France, Report 2: Image and marketing strategy. Division of Agro-Marketing, University of Ghent.

Viaene, J. and Gellynck, X. (1995) Doorlichting van de Belgische slachthuizen en uitsnijderijen: Varkensvlees, Deel I: Financiële analyse. Division of Agro-Marketing, University of Ghent., pp. 35–9.

6 Sophisticated consumers and export success, but problems in the home retail sector: the Italian pasta industry

LUCIANO VENTURINI AND
STEFANO BOCCALETTI

6.1 Introduction

The objective of this chapter is to analyse the determinants of international competitiveness and the sources of competitive advantages of the Italian pasta industry. This sector plays an important role in the Italian food industry and is generally considered a success story.

As a theoretical framework in assessing the competitiveness of this industry, we use the approach known as Porter's diamond. This approach, as indicated in Chapter 1, provides a conceptual framework to identify the factors and variables which determine the competitive performance of firms, countries and industries. It is based on the idea that the main source of competitive advantage has shifted from the determinants of static efficiency to the resources and intangible assets relevant to dynamic efficiency. The capacity to export and to develop the process of internationalization through foreign direct investment (FDI) is seen as related to the firm's home diamond.

This view has also received some criticism. Some authors have pointed out its excessive reference to the home base. In smaller countries a 'double diamond' framework has been proposed as a better approach, since in these countries firms may choose strategies not based on the home-country diamond alone.

In addition, some problems emerge when the framework is applied to analyse the competitiveness of multinational enterprises (MNEs). Porter's approach holds when an MNE builds only on the country-specific attributes of its home diamond to achieve international competitiveness. This view is challenged when the competitive position of MNEs, particularly the large ones, does not necessarily depend only on the characteristics of their home diamond. Increasing globalization and interdependence and the growing cross-border networking of firms through strategic alliances and other international co-operative arrangements undermine any concept of national firm-specific diamonds (Dunning, 1993; Rugman and Verbeke, 1993).

This critical literature provides useful arguments for modification and extension of Porter's diamond. One main implication is that national diamonds have to be replaced by supranational diamonds (Dunning, 1993).

However, the principle of the diamond, as Dunning admits, may still hold well in order to examine and evaluate the origins of competitive advantages of firms, industrial sectors and countries.

In particular, the role of the home diamond should be regarded as central in the case of firms and sectors whose degree of international involvement is not yet highly developed in terms of FDI and co-operative arrangements. At least during the first stages of the internationalization process, the nature and characteristics of the home diamond remain crucial. The home diamond remains the appropriate framework to examine whether and how domestic firms develop a competitive advantage in these initial stages. As we shall see, Porter's diamond provides a useful theoretical framework to analyse the determinants of international performance of the Italian pasta industry. This seems consistent with the idea of Dunning (1993) that the pattern of the diamonds of countries will differ according to the extent and form of their involvement in the global economy.

The remainder of this chapter is organized as follows: section 6.2 provides an analysis of the international performance of the industry, in section 6.3 we apply the diamond framework in order to explain the facts detailed in previous sections, and a summary and evaluation of the explanatory power of Porter's framework for the industry examined is presented in section 6.4.

6.2 Export performance and international competitiveness

In this section we provide some measures of the international performance of the Italian food industry and of the pasta industry. The measures can be divided into two main categories:

- those which are conceptually based on trade measures
- those which include or take into account the role of FDI.

6.2.1 Trade measures

Conventional trade measures of performance signal clearly quite a good performance of the Italian pasta industry. Table 6.1 shows the level of revealed comparative advantage (RCA; see Chapter 1) of the industry. Both in volume and value the Italian index is very high, and increasing. No other European country performs in a similar way.

The same picture is provided by assessing the level of pasta exports as a percentage of domestic output (Table 6.2). The percentage for Italy was 22.4 in 1987 and increased to 31.2 in 1991. Every such indicator shows that the Italian performance is quite unique in Europe. It should be noted that the average propensity to export of the Italian food industry as a whole was 6.8% in 1983 and increased to 9.6% in 1992 (Pieri and Venturini, 1995). This means that the pasta industry export performance is quite high relative to the

Table 6.1 Revealed comparative advantage for pasta products in the EU countries: volume and value (indicated in bold)

	1988	1989	1990	1991	1992
Benelux	45	43	60	60	49
	69	**65**	**88**	**89**	**78**
Denmark	5	10	10	14	17
	10	**22**	**20**	**32**	**43**
France	40	36	36	32	28
	37	**34**	**36**	**32**	**31**
Germany	13	11	12	10	12
	19	**16**	**19**	**17**	**15**
Greece	64	74	65	60	57
	42	**47**	**39**	**41**	**41**
Ireland	20	19	18	12	9
	59	**61**	**56**	**38**	**25**
Italy	652	666	660	688	702
	597	**614**	**596**	**622**	**634**
Netherlands	18	16	16	13	12
	34	**32**	**32**	**24**	**25**
Portugal	14	7	7	10	4
	11	**7**	**8**	**12**	**8**
Spain	12	11	14	13	15
	15	**15**	**20**	**19**	**23**
UK	11	9	11	11	9
	20	**16**	**17**	**19**	**17**

Source: Pitts, 1996.

Table 6.2 Pasta exports as a percentage of domestic output (UK, value; all other countries, volume)

	1987	1988	1989	1990	1991
France	7.0	7.2	7.5	8.1	na
Germany	7.9	7.9	7.6	8.3	8.8
Italy	22.4	23.4	26.8	26.8	31.2
Spain	0.5	0.6	1.2	2.5	3.1
UK	0.8	0.6	0.8	0.8	1.2

Source: Market Research Europe, 1994.

average food performance. This is also confirmed by data (not reported) on net exports.

Looking at the relative position on the international pasta market, we see that Italy is, by a large margin, the main exporter of dry pasta in the EU with 435 120 tonnes in 1995 and the main exporter on the world market with 976 460 tonnes in the same year. Export flows of Italian dry pasta during the period 1988–1995 show a continuous growth: in volume, they more than doubled (Table 6.3). The contribution of EU markets to the growth was substantial up to 1994. Only in 1995 was there a slight reduction of the quantity marketed in the EU-12, while for the EU-15 the data do not show

Table 6.3 Export flows of Italian pasta: volumes by main destination countries (tonnes)

	1988	1989	1990	1991	1992	1993	1994	1995
Belgium-Lux	14 805	14 609	15 386	16 630	16 829	19 562	20 662	21 231
Denmark	6 927	7 381	9 391	9 964	11 743	14 586	15 537	17 768
France	85 310	83 221	92 445	100 683	112 600	125 704	133 844	129 347
Germany	67 675	72 919	79 570	98 311	110 289	139 282	143 518	141 414
Greece	3 241	4 724	6 385	10 384	10 011	13 158	15 743	13 542
Ireland	593	729	692	867	1 331	970	1 498	1 988
Netherlands	11 093	12 760	14 233	15 973	15 149	18 839	20 720	22 277
Portugal	981	1 761	1 559	2 179	2 406	3 207	4 321	4 402
Spain	3 204	3 111	3 285	2 651	2 447	3 653	5 371	5 663
UK	40 198	52 594	48 883	56 753	67 163	73 939	79 773	77 489
EU-12	234 026	253 809	271 828	314 395	349 966	412 899	440 987	435 120
EU-15	247 767	272 367	295 671	343 592	383 044	448 598	468 349	486 679
Central-Eastern Europe	397	926	14 578	97 780	259 247	125 329	77 936	84 737
US	56 014	80 582	92 908	79 717	90 051	111 363	140 811	153 334
Canada	10 957	11 295	11 672	12 478	12 348	18 977	20 148	19 443
Japan	36 698	42 885	38 474	43 670	44 825	50 856	48 926	59 764
Total	433 146	515 854	545 267	688 605	905 163	904 826	920 445	976 460

Source: Elaboration on ISTAT data.

any slackening of the increasing trend. On the other hand, the main destination country for Italian pasta, the US, in 1995 absorbed a volume of product almost three times larger than in 1988, representing 16% of the Italian total pasta exports. It therefore represents a strategic target for exporting pasta producers (Table 6.4).

About 45% of total exports goes to the EU-12 market, 50% considering the EU-15; the other half goes to the already mentioned US market, and to other markets, the most important of which is Japan, with a relatively steady share of 5–6% of total exports. The strategic role of Japan is confirmed by the joint-venture developed by the market leader Barilla with a local producer, Ohmi Food (Quarleri, 1994). The countries of central and eastern Europe represent an important market, which grew spectacularly after 1989. Recent political developments have weakened the financial position of these countries, with important effects also on the import demand for pasta, as can be inferred from Table 6.4. Their demand reached a peak in 1992, with a share of 29% of total Italian pasta exports, declining to a level of 8–9% in subsequent years.

Within the EU-12 (Table 6.5), traditional markets have probably reached market saturation, with a stagnant demand for the Italian product: in particular, the share of France (which was the most important European market for Italy until 1992, when it was replaced by Germany), in total Italian exports to EU-12 diminished from 41% in 1985 to a steady 30% for the three years 1993–1995. Neither Germany nor the UK experienced the same reduction, although their shares stabilized at 33% and 18% respectively in 1994–95. In absolute terms, the role of these three main countries has increased over time, the only exception being 1995.

Table 6.4 Export flows of Italian dry pasta: percentage of volume by destination

	1988	1989	1990	1991	1992	1993	1994	1995
Belgium-Lux	3	3	2	2	2	2	2	2
Denmark	2	1	2	1	1	2	2	2
France	20	16	17	15	12	14	15	13
Germany	16	14	15	14	12	15	16	14
Greece	1	1	1	2	1	1	2	1
Ireland	0	0	0	0	0	0	0	0
Netherlands	3	2	3	2	2	2	2	2
Portugal	0	0	0	0	0	0	0	0
Spain	1	1	1	0	0	0	1	1
UK	9	10	9	8	7	8	9	8
EU-12	54	49	50	46	39	46	48	45
EU-15	57	53	54	50	42	50	51	50
Central-Eastern Europe	0	0	3	14	29	14	8	9
US	13	16	17	12	10	12	15	16
Canada	3	2	2	2	1	2	2	2
Japan	8	8	7	6	5	6	5	6
Other	10	21	17	16	13	16	19	17
Total	100	100	100	100	100	100	100	100

Source: Elaboration on ISTAT data.

Table 6.5 Export flows of Italian dry pasta in the EU-12: percentage volumes by destination

	1984	1985	1986	1987	1988	1989	1990	1991	1992	1993	1994	1995
Belgium-Lux	8	7	6	6	6	6	6	5	5	5	5	5
Denmark	2	2	2	2	3	3	3	3	3	4	4	4
France	39	41	40	40	36	33	34	32	32	30	30	30
Germany	30	27	26	27	29	29	29	31	32	34	33	33
Greece	2	2	1	1	1	2	2	3	3	3	4	3
Ireland	0	0	0	0	0	0	0	0	0	0	0	0
Netherlands	5	5	6	4	5	5	5	5	4	5	5	5
Portugal	NA	NA	0	0	0	1	1	1	1	1	1	1
Spain	NA	NA	1	2	1	1	1	1	1	1	1	1
UK	15	17	17	17	17	21	18	18	19	18	18	18
EU-12	100	100	100	100	100	100	100	100	100	100	100	100

Source: Elaboration on ISTAT data.

The relative loss of importance of the French market can be explained by looking at the share of the Italian product on the total import of dry pasta (egg pasta excluded) in France: Italy is by a long way the market leader with a share over 90%, though this figure declined to less than 80% in 1994 (Table 6.6). Exports of Italian non-egg dry pasta to France experienced a two-year decline, with the volume exported falling from more than 80 000 tonnes in 1992 to about 69 000 in 1994. In 1993 the Italian losses were hidden by a decrease in total French imports, but the share was affected heavily in 1994 after an increase of total imports from 82 000 tonnes in 1993 to over 86 000 in 1994. The loss of competitiveness of the Italian product has played in favour of Belgian pasta, which increased its export flows to France from 1500 tonnes in 1993 to more than 10 000 tonnes in 1994. This is not the result of a re-exporting activity of Italian pasta from Belgium: in fact the export flow from Italy towards this country, about 14 000 tonnes, did not show significant changes in 1994.

The other markets where the Italian product seemed to lose competitiveness are Portugal and Spain; for these two countries the trend can be explained by the fact that Barilla, the largest Italian company, acquired two plants in Spain in 1988: output from these plants may have replaced the product originating from Italy. The Italian market share has improved in Denmark and Netherlands, and remains quite steady (at a very high level) in Germany, Greece and the UK.

In the EU-12, over the period 1988–1994, after the complete opening of national markets to pasta products even with a percentage of soft wheat, the Italian share of total imports decreased from 83.0% to 78.9%. Nevertheless, the absolute level of EU imports of Italian pasta is increasing, indicating that the growth of the demand for pasta has been fulfilled substantially by greater exports from other countries.

In particular, the Belgian pasta industry has assumed an important role in the EU, accounting for 8.2% of total intra-EU imports in 1994 (Table

Table 6.6 Italian dry pasta as a percentage of total imports of pasta (egg pasta excluded) for the main EU markets

	1988	1989	1990	1991	1992	1993	1994
Belgium-Lux	55.1	57.0	46.4	45.7	49.4	36.7	42.6
Denmark	72.3	61.4	85.2	79.0	78.7	86.3	84.7
France	92.7	92.4	91.8	92.9	93.9	91.4	79.4
Germany	93.3	93.6	92.4	92.1	90.7	92.1	90.5
Greece	99.3	98.3	99.9	99.1	98.1	97.4	98.5
Ireland	35.7	40.0	28.0	39.1	39.9	8.6	29.7
Netherlands	44.6	51.8	51.7	53.4	52.8	55.3	57.8
Portugal	83.0	81.8	80.1	79.2	84.9	50.9	65.1
Spain	96.8	85.0	80.7	75.9	62.9	53.8	72.2
UK	85.8	87.3	86.2	88.6	87.3	83.7	89.5
EC-12	83.0	83.9	82.4	82.9	83.2	80.5	78.9

Source: Elaboration on ISTAT data.

Table 6.7 Shares of main European pasta exporters on intra EU-12 imports

	1988	1989	1990	1991	1992	1993	1994
Italy	85.5	86.4	84.5	85.0	85.2	82.9	81.5
Belgium	4.2	4.0	5.1	5.2	5.3	5.4	8.2
France	3.5	3.2	3.0	2.8	2.4	2.5	2.7
Germany	2.6	2.5	4.0	3.6	2.8	3.1	2.8
Greece	1.8	1.3	0.8	0.7	0.6	0.5	0.5
Spain	0.1	0.5	0.5	0.7	0.8	2.2	1.5
Others	2.3	2.1	2.1	2.0	2.9	3.4	2.8
EEC-12	100.0	100.0	100.0	100.0	100.0	100.0	100.0

Source: Elaboration on ISTAT data.

6.7); in fact, the reduction of the Italian share of total intra-EU import, four percentage points between 1988 and 1994, has been transferred mostly to Belgium. Among the other main exporters, France and Germany seem to have stabilized at slightly above 2.5%, whereas Greece has lost ground, accounting for only 0.5% of total imports. The increasing role of Spain in recent years is mainly due to the already mentioned investments by Barilla.

It is hard to say if the small loss of competitiveness of the Italian product is due to institutional reasons, linked to the adoption of the principle of mutual recognition of national standards in the EU, initiated by the White paper of the European Commission on the creation of the single market and the subsequent decision of the Court of Justice (14 July 1988), or to the diminished efficiency relative to other exporting countries. The data on comparative advantage seem to exclude any efficiency problem (Table 6.1).

Before the decision of the European Court, a 1967 Italian law barred any import of dried semolina pasta produced with even a small percentage of soft wheat. Since 1988 it has been possible to market these types of pasta in Italy. On the European market Italian pasta produced in the traditional way is sold with the same labelling as pasta produced at least partially with soft wheat, thereby creating a hypothetical confusion in the minds of consumers. So far, the fears of Italian producers regarding the domestic market seem to be groundless; on the other hand, even if the Italian product is still competitive on the EU market, the data illustrated above seem to indicate the necessity for strategies to counteract the emerging negative trends.

An interesting case is France, which had imposed the same type of legislation on its producers, forcing them to use only hard wheat in pasta products. Pasta imports have increased since 1990, and with it the appearance of new suppliers, such as Belgium and the Netherlands. Overall, in the EU-12 Belgium is emerging as an aggressive competitor in the pasta market (Table 6.7).

6.2.2 The role of FDI

In an increasingly integrated world economy, the concept and the measures of international performance and competitiveness become more complex. Traditional indicators are no longer sufficient to measure overall international performance. A complete picture of international competitiveness should not be seen in terms of trade alone but should consider the foreign investment activity of national firms.

Measures of competitiveness which take into account the foreign production of the firms of a country may be more useful than single trade based measures (Traill and Gomes da Silva, 1996). Of course, the more interesting issues emerge when different measures provide different pictures. Generally, these differences illustrate that national firms may be competitive globally even when the domestic based production seems weak in terms of market share (Kravis and Lipsey, 1992). In our case, it is interesting to explore the hypothesis that, even though traditional trade measures indicate a good export performance, there might be problems in terms of more complete measures of the internationalization process.

As already noted by Porter, a weakness in internationalization is a general characteristic of Italian firms:

> while successful Italian firms are extremely international in outlook, foreign direct investment is comparatively rare. The Italian position abroad is largely won through exports. (Porter, 1990)

Table 6.8 confirms the weak position of Italy as a direct investor. Over the period 1987–1992, the Italian rate of FDI as a percentage of GDP was very

Table 6.8 Foreign direct investment flows (as % of GDP)

	Inward investment		Outward investment	
	1980–86	1987–92	1980–86	1987–92
Belgium	1.15	3.85	0.36	3.30
Denmark	0.16	0.73	0.35	1.23
France	0.39	0.86	0.53	1.60
Germany	0.12	0.25	0.64	1.16
Greece	1.35	1.52	n.a.	n.a.
Ireland	0.85	0.24	n.a.	n.a.
Italy	0.20	0.45	0.38	0.51
Netherlands	0.79	2.04	2.02	3.82
Portugal	0.84	2.92	0.05	0.35
Spain	1.07	2.00	0.17	0.41
UK	1.14	2.51	2.13	3.03
EC-12	0.53	1.16	0.85	1.55
United States	0.59	0.84	0.30	0.53
Japan	0.03	0.03	0.44	1.09
OECD countries	0.46	0.82	0.56	1.11

Source: OECD, 1994.

low: within the EU, only that of Spain was lower. Indeed, this also is true of the Italian food industry. The cumulative balance of outward investments is smaller than the cumulative balance of inward investments over the period 1983–1991 both for manufacturing industries and for the food industry (Table 6.9). For outward investments by food companies, in 1992 and 1993 the trend seems reversed, with a value approximately three times larger than before, which may indicate a late opening of Italian firms towards international markets, probably stemming from the higher degree of market concentration, which developed especially in the 1990s.

In terms of mergers and acquisitions in the world food industry over the period 1989–1993, Italy is not one of the top investing countries, but it is included in the top target countries (Table 6.10).

Traill and Gomes da Silva (1996) analysed the competitiveness of the Italian food industry using traditional measures (RCA) and also newer measures which took account of the degree of FDI. They found that the revealed comparative advantage measure for the Italian food industries shows a slight upward trend, while the trend in net exports was slightly downwards. This means that Italian-based food firms are competing relatively better than the other Italian industries in world markets.

However, when foreign production is taken into account, the indices indicate that Italian firms are losing competitiveness to their foreign rivals. Such a negative international performance is typical of Italy. French food firms are doing well internationally, German food companies are improving their performance relative to the rest of the world's food companies and British firms are relatively strong performers.

We argue that the Italian pasta industry also appears less competitive when the overall performance of its firms is considered. The problem of data availability on FDI at a disaggregated level precludes an appropriate quantitative analysis. However, qualitative analyses and case studies show that, despite the success of the industry in terms of traditional trade measures, its internationalization process in terms of FDI seems to be late and rather slow (Pieri and Venturini, 1995; Eurofood, 1996a).

The weakness of the internationalization process of Barilla, the leading firm in the Italian pasta industry, gives some clear indications here. (As we shall see in section 6.3, Barilla has a dominant role in the pasta sector in Italy, followed by a limited number of medium-sized firms and a fringe of small local producers.) Given the association between firm size and the level of FDI, the internationalization process of the sector in terms of FDI is substantially determined by the sector leader's foreign activities. From this point of view, notwithstanding its strong role as an exporter, Barilla's internationalization is relatively weak. Table 6.11 shows, for example, that Barilla has a rather small number of employees abroad (less than 1000). In 1988, the share of Barilla's total employment abroad was only 7.6%, with a jump between 1990 and 1991 due to the acquisition of the Greek leader Misko.

Table 6.9 Foreign direct investment flows for Italy (billion lire): Food industry and total manufacturing industry

	1983	1984	1985	1986	1987	1988	1989	1990	1991	1992	1993
Inward investments											
Food industry											
Investments	83	134	246	95	301	707	480	866	136	585	300
Disinvestments	23	40	46	76	134	166	618	93	1	18	43
Balance	60	94	200	19	167	541	-138	773	135	567	257
Manufacturing industry											
Investments	1937	1798	2080	3074	2775	6633	6017	8877	6147	3010	3600
Disinvestments	558	321	1714	4988	579	2589	4097	8401	3459	1151	846
Balance	1379	1477	366	-1914	2196	4044	1920	476	2688	1859	2754
Total net investments	1807	2267	1916	-23	5264	8902	3469	7569	3152	3945	5627
Outward investments											
Food industry											
Investments	33	18	81	74	114	35	283	456	339	1372	1325
Disinvestments	69	2	7	7	39	341	88	143	269	402	304
Balance	-36	16	74	67	75	-306	195	313	70	970	1021
Manufacturing industry											
Investments	1333	993	908	1415	1999	1338	3604	3795	4712	8310	7139
Disinvestments	115	125	590	133	871	942	3312	1832	2575	7389	3378
Balance	1218	868	318	1282	1128	396	292	1963	2137	921	3761
Total net investments	-3230	-3505	-3471	-3968	-3017	-7094	-2748	-8682	-8277	-6929	-10 945

Source: Banca d'Italia, Relazione annuale, several issues.

Table 6.10 Merger and acquisition activity in the world food industry, 1989–1993

Most active investors by number of deals	No. of deals	$(million)	Top investing countries by value	No. of deals	$(million)	Top target countries by value	No. of deals	$(million)
Unilever	63	1830	US	273	15 079	UK	154	10 179
BSN	55	3464	UK	278	8 310	US	190	7 463
Nestlé	38	4129	France	215	6 211	France	133	4 948
Grand Met	37	900	Switzerland	81	4 457	Switzerland	9	3 821
Pepsi Co	34	2401	Japan	109	2 566	Italy	60	3 520
			Other	745	16 543	Other	1155	23 235
			Total	1701	53 166	Total	1701	53 166

Source: Eurofood, 1994.

Table 6.11 Number of total and foreign employees of Barilla

	1988	1989	1990	1991[a]	1992
Employees in foreign plants (1)	407	386	348	919	885
Total employees (2)	5326	5994	6045	6776	8252
(1)/(2) (%)	7.6	6.4	5.8	13.6	10.7

[a] Acquisition of the Greek leader Misko.
Source: Mediobanca, 1994.

6.3   The determinants of industry's competitiveness

According to Porter (see Chapter 1), the international success of an industry depends on the existence of four well-known factors or sources of competitive advantage:

- factor conditions
- demand conditions
- related and supporting industries
- firm strategy, structure, and rivalry.

These four basic factors create a **home diamond**. Nations are most likely to succeed in industries where the home diamond is most favourable in the sense that it pressurizes and supports the national firms in investing in specialized assets and skills. Nations whose home environment is the most dynamic and challenging succeed, since their firms are stimulated to upgrade and widen their advantages over time.

A crucial point of Porter's approach is the idea that the diamond is a mutually reinforcing system. Advantages in one determinant can also create or upgrade advantages in others. Such an interaction among the determinants yields self-reinforcing benefits. The effect of one source of competitive advantage depends on the state of others. Favourable demand conditions, for example, will not lead to competitive advantage unless domestic rivalry is sufficient to cause firms to respond to it.

This means that though advantage in every determinant is not required for competitive advantage in an industry, the competitive position is stronger the more numerous are the interacting sources of competitive advantage. If the competitive advantage is based only on one or two determinants, it usually proves unsustainable.

The conceptual framework developed by Porter (1990) seems quite appropriate to analyse the determinants of export performance of the Italian pasta industry. It can simultaneously explain both its good performance in terms of exports and some weakness in its internationalization process.

In what follows, we show that the specific nature of the Italian home diamond plays a crucial role in determining the strengths and weakness in the competitive position of the Italian pasta firms. At the same time, Porter's original framework provides an appropriate tool in explaining the international performance of this industry.

We argue that the Italian pasta industry might have used only some factors or sources of competitive advantage, while the domestic competitive environment was substantially deprived of other crucial determinants of competitive advantage. The limited availability of these sources and the impossibility of obtaining the self-reinforcing benefits by a complete interaction of more numerous factors seem crucial to an appropriate evaluation of the industry's performance.

6.3.1 Factor conditions

Factors of production, that is the inputs necessary to compete (such as the quantity, skills and cost of human resources; the abundance, quality and cost of physical resources; the stock of scientific, technical and market knowledge; the amount and cost of capital resources to finance industry; and the type and user cost of the nation's infrastructure) may positively contribute to creating competitive advantage if a country's firms possess them at low cost or can get unique high-quality factors that are relevant to competition in a particular industry.

While such a match between industries and the country's factors is the core of the standard theory of comparative advantage, Porter (1990) argued that the role of factor endowments is more complicated and that the mere availability of factors is not a sufficient condition to determine competitive success. Instead, this result depends on how efficiently and effectively the factors are deployed.

In this sense, Porter tends to de-emphasize the role played by factor endowments. This view seems particularly appropriate for more basic and less specific factors. Only advanced factors and man-made ones (such as modern communications infrastructure, highly educated personnel, university research institutes), particularly if they are specialized (i.e. relevant to specific industries), provide a decisive source for a sustainable competitive advantage.

A corollary of this view is that industries relatively more dependent on basic and generalized factors available in many countries may be successful only if a country can count on the positive influence of other parts of the diamond. This seems the case of our industry. The pasta industry does not need particularly advanced and specialized factors of production. As Porter himself notes, Italy in general draws relatively few advantages from created factors of production. It is relatively weak in formal research and remains characterized by little formal training; advanced training is generally informal and on-the-job.

Instead, a crucial and well-known role in the accumulation of knowledge is played in Italy by the geographic concentration of firms. The existence of

industrial districts (see Chapter 1) is the main source of competitive advantage for several Italian industries. In industrial districts, the physical proximity of suppliers and customers creates networks, which facilitate co-operation and specialization. The geographic concentration of firms reduces transaction costs and leads to a rapid accumulation and diffusion of knowledge.

The pasta industry, however, is not characterized by location patterns based on the industrial district (ID) model. Observation indicates that the clustering of firms in the pasta industry is not an important phenomenon. Pasta manufacturers are scattered and not geographically concentrated. The positive role that proximity played in other industries concentrated in industrial districts, for example the tomato industry, did not work in this case. Other factors have probably played a role in the success of Italian pasta products, such as the availability of domestic durum wheat and a technological advantage, built through a long tradition in pasta production. These advantages, which refer to a product of medium–high quality, allow the preservation of the leadership on the world market and in the main destination countries, where the market segments demand a high quality product.

In summary, the pasta industry does not require highly advanced and specialized knowledge and skills; on the other hand, the endowments of these factors and their creating mechanisms are not particularly well developed in the Italian environment. This conclusion leaves open the search for which other parts of the diamond might have played a crucial role in determining the competitive success of the industry.

6.3.2 Demand conditions

Among the factors at work in determining the Italian performance in the pasta industry, demand conditions play a positive and relevant role. Porter points out that three characteristics of home demand are particularly significant to achieving national competitive advantage: its size, the sophistication of consumers and the ability to anticipate changes in foreign demand.

The relative size of home demand may be relevant in the presence of significant static economies of scale or dynamic learning. However, more important than the size of home demand is the nature of home buyers. A crucial determinant of competitive advantage is the presence of sophisticated and demanding buyers. Their presence pressures local firms to meet high standards in terms of product quality, to move into newer and more advanced segments over time, and to improve service. The third important dimension is the possibility that home demand could anticipate those of other nations. In this case, home demand provides an early warning indicator of buyer needs.

Home demand has played a crucial role in determining the export performances of pasta products. There is no doubt that both the size of home demand and the degree of sophistication of the Italian consumers resulted in favourable demand conditions. As indicated in Table 6.12, the size of the domestic market is quite significant both in absolute terms and in comparison to the market size of other countries.

Equally relevant is the advantage deriving by the desire for quality. Italian consumers place a very high importance on pasta's quality, as exemplified by cooking characteristics, aspect, colour, taste, and aroma (Braga and Raffaelli, 1995). This is witnessed by the gradual decline of the importance of the low quality–low price segment of the pasta market and by the increase of the average quality level of pasta products (Torazza, 1990; Eurofood, 1995).

Given the size of the domestic market and the sophistication of final consumers, these demand conditions have clearly contributed to stimulate the evolution of the industry towards high quality standards and to develop an intangible asset in terms of good reputation. The reputation models in the literature (Bond, 1984) hypothesize that the cost structure of two different countries could give a comparative advantage in the production of low quality product to one and an advantage on high quality products to the other. This poses some problems with reference to emerging markets, where the mass market has not yet been educated to the consumption of 'good' pasta; in this case other factors will probably determine the success of a supplying country: economies of scale, first mover advantage, promotional activities, etc.

Finally, in the case of pasta, the ability to anticipate market developments does not depend on factors connected to the stage of development as implicitly hypothesized by Porter, who refers essentially to high technology products. Instead, this ability is essentially explained by the Italian eating habits, the country's national passion and specific taste for pasta: these factors are clearly unrelated to the stage of development and Italian consumers find themselves as 'early adopters' in a sense quite different to what happens in the case of high technology products.

The Italian pasta industry has certainly benefited from the increasing attention to diet and from the increasing perception of pasta as a convenience and

Table 6.12 Per capita consumption of pasta products in selected countries, 1992

	kg. per capita		kg. per capita
Italy	23.0	Denmark	2.0
Greece	8.5	UK	2.0
Portugal	7.0	Ireland	1.0
France	6.7	Venezuela	12.7
Germany	4.5	Switzerland	9.1
Belgium/Lux.	4.5	USA	8.5
Spain	4.1	ex Soviet Union	7.0
Netherlands	4.0	Argentina	6.8
Austria	3.9	Egypt	6.8
Sweden	3.5	Canada	6.3

Source: Databank, 1995.

healthy food product in most advanced countries. In these countries pasta tends to become a convenient course, with its ease and speed of preparation. As a low-fat, low-cholesterol and low-calorie food, pasta has become increasingly popular with health and fitness-conscious consumers throughout the world. In the eight major world markets the variety of pasta available reflects growing consumer sophistication. In particular, the fresh pasta sector is by far the most dynamic, benefiting from the shift towards premium products and improved packaging. Industry analysts predict that every country apart from Germany and Italy will register increasing pasta sales over the next 5 years, as the health-awareness trend continues (Eurofood, 1996b).

Italian consumers are 'early adopters' of this trend and Italian industry benefits from this specific, exogenous and unexpected change in food tastes, strongly influenced by diet and health attention. This exogenous change should be considered as a chance event, since it is outside the control of the Italian firms. Obviously, these major shifts in foreign market demand offered a unique opportunity to Italian firms.

An analogous positive contribution has not, however, been provided by retailer demand. Porter recognizes that the role of sophisticated and demanding buyers can be played by distribution channels as well as final consumers. The presence of large powerful chains stimulates producers to cut costs, to create new customer services and to launch new products and varieties.

In this way, Porter (1990) extends the potential role which might be played by buyers and retailers in determining the competitiveness of their suppliers. In his earlier work (see for example Porter, 1980) he focused essentially on the possibility that strong buyers and retailers used their bargaining power to bargain away the profits for themselves. This hypothesis is well rooted on the conventional approach of the structure–conduct–performance paradigm with its typical emphasis on profitability levels, arising from a firm's market power, as the main dimension of performance. According to this view, strong retailers lower manufacturers' profitability. The recent work of Porter extends the analysis on the potential role of retailers by considering their impact on dynamic efficiency. Porter (1990) provides anecdotal and empirical evidence of this positive role of retailers. In addition to the case of eyewear store chains, which contribute positively to the competitiveness of the US contact lens producers, Porter indicates the cases of some Italian industries:

> (in Italy), such products as shoes, clothing, furniture, and lighting are sold in greater proportion through specialty stores than in other nations. These sophisticated retailers are a major force pressuring Italian manufacturers to constantly introduce new models and reduce prices and thus costs. (Porter, 1990, p. 90)

In addition, in his analysis of the Italian demand conditions, Porter (1990) observes that the sophistication of final consumers in Italy is reinforced by the presence of sophisticated distribution channels. In his words in areas such as clothing, footwear, tiles and furniture . . . Italian retailers are almost invariably smaller and more specialized in particular products than foreign retailers. They are intimately familiar with their business and represent an extremely knowledgeable and demanding intermediate buyer for the products involved. Italian firms must constantly come up with new models to secure and maintain distribution.

Such forces were not at work in the case of the pasta industry or for most Italian food industries, given the fragmented structure and unsophisticated nature of the Italian grocery retailing which is characterized by a significant number of traditional independent small retailers. (Some evidence about this fragmentation of the retailing sector in Italy is provided later in the chapter.) For the moment we emphasize that the existing structure has impeded distribution channels from reinforcing the positive contribution represented by sophisticated final consumers. It should be noted that in grocery retailing, more than in other sectors of retailing, the retailers' degree of sophistication is strongly related to their size and to the industry's structure. In other words, the small size of firms and the low concentration of the industry result in unsophisticated buyers.

Until quite recently the structure and organization of the Italian retailing sector has exerted a double negative effect on the upgrading of competitive advantage well focused by Porter (1990). They have resulted in elevated costs and low labour productivity in retailing and, even more important, they have blocked the development of advanced retailers and sophisticated channels in Italy. This absence has deprived the Italian pasta industry of a relevant factor in upgrading competitive advantage.

6.3.3 Related and supporting industries

The third broad determinant of national advantage in Porter's diamond concerns the presence of internationally competitive supplying industries or related industries. This presence creates several benefits and potential advantages in downstream industries, for example in terms of the availability of efficient, early, or rapid access to cost-effective inputs. This facilitates the process of innovation and upgrading. Close working relationships with sophisticated suppliers help firms to apply new technologies and allow quick access to information, new ideas and insights. Further, firms have the opportunity to influence suppliers' technical efforts. The co-ordination of R&D and joint problem-solving improve and increase the pace of innovation. In this case, geographical proximity may be important. All the indicated benefits are enhanced if suppliers are located in proximity to firms, since proximity helps the process of interaction and shortens the communication lines.

In this regard, the Italian pasta manufacturers seem supported by a positive environment. The domestic sector of machinery for pasta and cereal processing, in particular, is quite successful internationally. Exports are more than 40% of total sales and it is a clear case of an industry based on sophisticated interfirm relationships, as a result of vertical disintegration of the production phases, and the development of networks among firms. The industry, being essentially characterized by the presence of SMEs, is highly flexible, while the existence of networks and co-operative interfirm relationships guarantees good performances in terms of product and process innovation. As a result, Italian technology is placed at a good level internationally in an industry which seems still characterized by some technological opportunities.

The most important innovation in recent years has been the use of high temperature drying in the pasta making process: it allows a better cooking quality, which is the most important parameter taken into account by the average consumer. The problem is the secondary effect of such a treatment: a lower nutritional value of the product. As a result this innovation is applied by a small number of firms, catering for the lower end of the market, while those with products perceived as high in quality often prefer to achieve the necessary standard of cooking quality by alternative means, such as using high quality durum wheat and optimizing the entire distribution channel, thus preserving the nutritional value of their products.

For example De Cecco, one of the leading high quality producers, developed its own processing system, with a maximum drying temperature of 70°C. The market leader Barilla controls the quality of raw materials first by drawing a map of qualities for the durum wheat, so that only 15 days after the harvest they know to a good approximation the quality difference among the different geographical areas. Secondly, they check the quality of wheat stocks silo by silo. Moreover, 65% of the semolina is produced in mills integrated with the company.

The new technologies available have increased the minimum efficient scale of the plants, so that medium and large companies are affected by two opposing tendencies: on the one hand, the adoption of the new technologies allows them to exploit scale economies and therefore to reduce costs; on the other hand, the use of more traditional technologies assures a higher quality standard that, with adequate information to consumers, may reward the company with the higher prices which flow from product differentiation and reputation.

At this point, it is hard to say what tendency will prevail: the exaggerated attention of consumers to prices and increasing market concentration seem to promote the adoption of the new production technologies. In any case, it is also true that further innovation towards the development of *ad hoc* production systems, as in the case of De Cecco, may solve the dilemma. Moreover, the evolution of technologies involving electronics, particularly linked with an extensive use of information technologies, such as plant and machinery automation, will probably push the pasta industry towards a higher degree of concentration.

The Italian pasta machinery industry has supported pasta manufacturers with a flow of new technologies. As we have seen, the creation of networks of firms and the lack of vertical integration in the industry allow the advantages of small size flexibility, while exploiting the economies of scale and scope.

According to some industry analysts, the pasta processing sector presents some problems which might reveal points of weakness as the environment in the European market becomes more and more competitive. The main problem seems due to the heterogeneity of firms and the marked differences between them. The share of SMEs may be excessive and mean a delay in the concentration process. This might explain the fact that the process of internationalization is still based mainly on exports rather than on more modern and direct forms of presence abroad. Only a limited number of firms formed their own commercial branches abroad or became involved in joint ventures with foreign partners. Rolfo *et al.* (1993) point out that this small number of joint ventures may indicate some difficulties in competition against companies endowed with technologies which are as advanced or at least comparable to those available in Italy. Their conclusion is that only an improvement in the structure of the sector as a whole will make it possible to continue to successfully face foreign competitors and maintain the traditional performance of excellence. In any case, despite the probable existence of some factors of weakness, there is no doubt that the Italian pasta machinery industry has performed quite well up to now and provided appropriate support to pasta manufacturers. As a supporting industry it has played a positive role in maintaining the competitiveness of pasta products.

Finally, it should be noted that Italy has a generally weak position in services: the only exception is design services. These services are important, and indeed they have contributed to the performances of many successful Italian industries (such as clothing, jewellery, furniture, automobiles). They are clearly less relevant for pasta products, and they have had only a minor positive influence on the pasta industry.

6.3.4   Domestic structure and rivalry

Porter's results provide empirical support for the hypothesis of an association between vigorous rivalry of domestic competitors and the creation and persistence of competitive advantage. A number of strong local rivals is necessary to establish a leading competitive position. Successful firms which compete vigorously at home pressure each other to improve and innovate. In this regard, according to Porter, a large home demand does not *per se* lead to competitive advantage. Without strong domestic rivalry, a large home market may induce complacency rather than stimulate investment and force firms to export and develop the internationalization process.

This seems to be the case with the Italian pasta industry. The market structure of the industry has remained rather fragmented until recently. Non-price competition was very weak. Table 6.14 shows that at the beginning of the 1980s the industry was still characterized by a low advertising intensity.

The advertising to sales ratio was only 0.5% in 1981 and 1.3% in 1986. Market concentration, as measured by concentration ratios, was also low. CR4 was only 30.2% in 1981; Barilla, with a market share of 20.0%, was the industry leader. The followers were much smaller firms, with market shares around 3–4% each (Table 6.15).

As a consequence, domestic rivalry was not at all vigorous until quite recently and its role as a pressing force to improve and innovate was rather weak. We argue that this not very competitive environment was also the result of the absence of strong retailers. We have already noted the negative consequences of this absence for creating appropriate demand conditions. However, the mechanisms through which retailers play a role do not work only on the demand side. They can also exercise a crucial influence on domestic rivalry. This happens through the impact of their countervailing power and, more directly, through the development of vertical competition as a consequence of private-label programmes (Venturini, 1994; Galizzi and Venturini, 1996). While Porter recognizes the potential role of retailing in upgrading competitive advantages, his analysis does not provide an adequate specification of the mechanisms at work. Porter seems to suggest that demand conditions are the only channel through which retailers exert their influence. In this way, he tends to undervalue the role that retailers may play, particularly in industries such as food and other packaged consumer goods. Powerful retailers can have a crucial impact even on other parts of the diamond, such as domestic rivalry. They can have a direct influence on the nature of competition (i.e. on the intensity of non-price competition) and the intensity of rivalry.

In a competitive environment characterized by a concentrated retailing sector, manufacturers face stronger incentives to innovate and differentiate their products. As a result, non-price competition becomes more intense and food manufacturers have to sink more resources in R&D, advertising and other marketing fixed costs. In other words, the increasing concentration of retailing and the high intensity of vertical competition accentuate horizontal competition among food manufacturers.

This possibility has been neglected by Porter, perhaps because of the broader nature of his analysis. Some crucial dimensions of vertical relationships, such as the phenomenon of vertical competition as a consequence of own label diffusion, can only be individuated and analysed by focusing on industries which sell mainly through channels of modern mass distribution. The food industry offers the opportunity of examining such more specific aspects and of noting how the presence of strong retailers with well developed private-label programmes implies more intense vertical and horizontal competition. Given the traditionally fragmented structure of retailing, the Italian case provides an opportunity to focus on these mechanisms.

The low levels of concentration and non-price competition in the Italian pasta industry may be at least partially explained by the absence of pressures and mechanisms which depend crucially on the concentration level of the retailing sector. The fragmentation of retailing is well illustrated by the retail distribution of pasta by outlet (Table 6.13). Hypermarkets and supermarkets control only 27% of total sales in Italy, a percentage much lower than that found in other countries. In Italy a higher percentage of pasta products is sold through traditional stores (47%). At the beginning of the 1980s, small traditional retailers still represented the dominant channel of distribution for pasta products, with 97.1% of total sales.

As a consequence of a fragmented retailing structure, pasta firms have not been pressured to increase their size, innovate and upgrade marketing skills. The leader did not face either strong horizontal competition or significant vertical competition from retailers.

Notwithstanding the process of structural change in Italian retailing in the 1980s, it seems clear that, until quite recently, Italian pasta manufacturers were not forced by strong retailers to adopt growth strategies and innovate. The presence of these pressures would have provided more stimulus both to the leader Barilla and to other competitors. The timing of Barilla's diversification would have been different. These strategies might have been adopted earlier. The same pressures might have induced other firms to adopt mergers and acquisitions, but for a long time these mechanisms did not operate or were too weak, contributing to a weak rivalry and little horizontal competition. The main driver of change was in fact the entry of MNEs at the end of the 1980s. In 1990 BSN (now Danone) took a major share in Agnesi. Its acquisitions of Italian pasta manufacturers continued in the following years, so that Danone is now the second largest operator in the Italian pasta market. In 1988, Nestlé acquired Buitoni, the second largest pasta manufacturer in Italy. These entries contributed to a radical change in the competitive environment.

Barilla reacted to the entry of MNEs by adopting a more aggressive growth strategy and increased advertising expenditures. Advertising intensity increased to 3.2% in 1993–94 from 0.5% in 1981 (Table 6.14). Finally, this signals a radical change in the nature of competition, which is now characterized by a significant emphasis on non-price competition and more intense horizontal competition. In particular, Barilla has been successful in develop-

Table 6.13 Retail distribution of pasta by outlet in 1992 (% volume; Italy: dry pasta only)

	Hypermarkets/ supermarkets	Small supermarkets/ Convenience stores	Traditional/ Speciality stores	Discounters	Others	Total
France	89	–	–	–	11	100
Germany	57	–	21	22	–	100
Italy	27	–	47	–	26	100
Spain	34	30	20	–	15	99
UK	65	15	–	–	20	100

Source: Market Research Europe, 1994.

Table 6.14 Advertising intensity in the Italian pasta market

	1981	1986	1993
Advertising expenditure (A) (billions of lire)	4.1	39.3	114.9
Sales (S) (billions of lire)	820.3	3097.6	3620.0
A/S (per cent)	0.5	1.3	3.2

Source: Largo Consumo, Databank, several issues.

ing aggressive non-price strategies on the domestic market by maintaining a 'share of voice' much larger than its share of market.

Market concentration has increased over the last decade as a consequence of the aggressive policy of the leader. As we have seen (Table 6.15), in 1981 the pasta industry had a four-firm sales concentration ratio of 30.2%. This ratio increased to 58.7% in 1995. Thus, during the late 1980s and in the 1990s the industry was characterized by a drastic process of concentration. The pattern of market shares indicates that the main change is due to the significant increase in Barilla's share. The market share of the leader increased from 20% in 1981 to 38.2% in 1995. The number of manufacturers has been reduced by half since 1964, although with a current total of about 160 the Italian pasta industry is still relatively fragmented (Market Research Europe, 1994).

The main determinant of the concentration process is the increasing intensity in non-price competition. Barilla has also improved its export performance. However, as we have already noted, the process of internationalization of the leading firm is still substantially limited to exports. Its FDIs were never significant and are still very low.

The picture which emerges from this analysis is that of a firm strongly involved in strengthening its position on a domestic market characterized by a changing competitive environment. In this context, less attention has been paid and few resources invested in more sophisticated strategies of internationalization.

Barilla's top management is quite revealing when, in this regard, they observe:

Table 6.15 Market shares and concentration ratios in the Italian pasta industry, 1981, 1989, 1995

	1981 Market shares	CR_n		1989 Market shares	CR_n		1995 Market shares	CR_n
1. Barilla	20.0		1. Barilla	25.0		1. Barilla	38.2	
2. Buitoni	4.0		2. Buitoni	6.0		2. De Cecco	9.2	
3. Amato	3.5	CR_3:27.5	3. Amato	5.5	CR_3:36.5	3. Agnesi	7.1	CR_3:54.5
4. Agnesi	2.7	CR_4:30.2	4. Agnesi	4.0	CR_4:40.5	4. Buitoni	4.2	CR_4:58.7
5. Garibaldo	1.3	CR_5:32.7	5. De Cecco	4.0	CR_5:44.5	5. Divella	4.0	CR_5:62.7

Source: Largo Consumo, Market Research Europe, Eurofood, several issues.

They say we should have expanded abroad earlier . . . our strategy was to strengthen ourselves at home first, considering the massive incursion of multinationals into the Italian market in the 1980s. Now internationalization is the priority. (Eurofood, 1996a)

The picture is clear. The entry of multinationals was an exogenous shock for the Italian leader. Its competitive position was satisfactory in the traditional configuration of the Italian market. But even the absolute and relative size, resources and capabilities of the leader were not adequate to pursue more sophisticated non-price strategies and to develop quickly the process of internationalization. Given the structural asymmetry and the fragmented nature of the Italian market, no other firm was able to develop a significant process of internationalization. Our hypothesis is that such a competitive position is the result of the absence in the Italian market of competitive pressures deriving from the retailing side.

A significant process of concentration has been observed over the last decade. When one analyses the 24-hour production potential of pasta factories over the period 1981–1991 (Table 6.16), it is clear that there has been a significant increase in the scale of plants. The increase in the minimum efficient size of the plants can be estimated as 170 tonnes per day. In 1991 24% of the units had a capacity greater than 100 tonnes per day; this share was 20.5% in 1986 and only 11.8% in 1981. Nevertheless, only 14 plants exceed 200 tonnes/day. The contribution of larger plants (i.e. exceeding 100 tonnes per day) to total production was about 70% in 1991, while the smallest ones (under 30 tonnes per day), which represent 50% of the total number of pasta plants, produce only 11.5% of total supply (Table 6.17).

Production is located especially in southern Italy, with 36.3% of pasta factories and 41.2% of total production; 32.2% of units are located in northern Italy, with a production share of 36.4%. The distribution of the plants is affected by factors such as the location of durum wheat production and the stronger demand for pasta products in the south, even though the increasing orientation towards foreign markets has caused a concentration of large pasta plants in northern regions, particularly in Emilia-Romagna.

A significant acceleration in non-price competition took place only as a reaction to the entry of foreign multinational companies in the mid-1980s. At this point, Barilla felt it necessary to reinforce its position through heavy advertising and promotional expenditure. This reaction has greatly contributed to change the competitive environment. The nature of competition increasingly emphasized non-price dimensions of competition. Advertising intensity and product innovation as well as the process of industry concentration increased. However, the crucial point is that the engine of change has been the entry of foreign competitors and the increased foreign ownership of the Italian industry. Even now, domestic rivalry is not particularly developed. Only a few firms have built national brands, and most pasta brands remain highly regional. Despite the process of concentration over the last

Table 6.16 Number of Italian pasta factories classified by 24 h production capacity and geographic area

Area	1981				1986				1991			
	over 100t	30.1–99.9t	up to 30t	Total	over 100t	30.1 99.9t	up to 30t	Total	over 100t	30.1– 99.9t	up to 30t	Total
North	9	13	49	71	15	6	16	37	17	7	31	55
Centre	5	8	25	38	7	7	13	27	6	8	13	27
South	14	18	54	86	15	16	38	69	17	19	26	62
Islands	0	12	31	43	1	13	18	32	1	11	15	27
Total	28	51	159	238	38	42	85	165	41	45	85	171

Source: UNIPI, 1992.

Table 6.17 Potential production of Italian pasta factories classified by 24 h production capacity and geographic area

Area	1981				1986				1991			
	Over 100t	30.1–99.9t	Up to 30t	Total	Over 100t	30.1 99.9t	Up to 30t	Total	Over 100t	30.1–99.9t	Up to 30t	Total
North	2010.0	777.0	674.0	3461.0	3307.0	340.0	574.0	4221.0	3538.0	370.0	551.5	4459.5
Centre	840.0	339.5	339.5	1519.0	1240.0	372.4	208.6	1799.4	1150.0	413.0	201.6	1764.6
South	2053.0	1040.8	724.4	3818.2	2984.0	959.0	521.7	4464.7	3604.0	1107.8	339.5	5051.3
Islands	0.0	627.0	535.0	1162.0	105.0	597.0	320.0	1022.0	160.0	517.0	316.4	993.4
Total	4903.0	2784.3	2272.9	9960.2	7636.0	2268.4	1624.3	11 507.1	8452.0	2407.8	1409.0	12 268.8

Source: UNIPI, 1992.

decade, the Italian pasta industry is still characterized by the presence of several small producers without sufficient resources and capabilities, particularly in the southern regions.

The European pasta industry has undergone considerable rationalization over the past couple of decades, with a sharp reduction in the number of operating companies and in the number of employees. According to the latest industry census, the number of manufacturers in France had fallen to 28 by 1984; 25 manufacturers operate in Germany; the Swiss market is supplied largely by 17 domestic factories with a mere 560 employees (Databank, 1995).

6.3.5 Chance

Porter defines **chance events** as

> occurrences that have little to do with circumstances in a nation and often largely outside the power of firms (and often the national government) to influence. (Porter, 1980, p. 124)

Among the main chance events, he indicates the surges of world or regional demand, in addition to pure invention, technological discontinuities, input costs, exchange rates, wars. The role of chance in influencing competitive advantage seems to have been quite significant in the case of our industry, because of the changes in international demand. A major shift in foreign market demand associated to the diffusion of the Mediterranean diet has played a crucial role in the case of pasta products. This shift created significant market opportunities for Italian pasta manufacturers and acted as a crucial determinant of export performance.

Although Italian firms clearly benefited from the diffusion of the Mediterranean diet, this eating style was not at all promoted by domestic players; in other terms, such an occurrence was beyond their control and should therefore be just considered as chance in Porter's sense.

6.3.6 The role of government

Porter's conceptual framework emphasizes the potential role of government in influencing the performances of domestic industries. Porter's hypothesis is that government is not a fifth determinant in addition to factor conditions, demand, related industries and domestic rivalry. Rather, government can play a role by influencing these four determinants either positively or negatively. In the case of many Italian food industries, including the pasta industry, this influence has been negative and operated mainly through regulation of the retail sector. This possibility is clearly predicted by Porter's theoretical framework and it is quite interesting to note that it has been emphasized in the case of Italy. Porter observed that:

> many nations have regulations that protect small independent wholesalers and retailers against the growth of large chains, or other restrictions on forming

modern distributive structures. While anti-chain store laws were abandoned decades ago in the United States, they persist in various forms in such nations such as Japan and Italy. (Porter, 1990, p. 650)

He continues by noting that such laws may serve social purposes or represent a form of protection from foreign competitors. But the crucial point is that restrictions on advanced distribution channels may, as we have seen, have negative effects on competitive advantage because they deprive the domestic environment of sophisticated and cutting-edge channels.

Since we argued that retailers may influence competitive advantage not only through their role as sophisticated buyers, but also through their impact on domestic rivalry, it should be noted that the negative effects of the Italian government's regulation in the retailing sector might have influenced even this part of the diamond. In other words, retailing regulations, by slowing or blocking the development of modern formats and channels of distribution, have not only deprived the Italian manufacturers of sophisticated buyers, but also impeded the intensification of domestic rivalry through vertical competition. These regulations have played a particularly negative role on the industry examined.

A second remark relates to government decisions. Public policies can have a positive impact on competitiveness. In some cases, regulation could be an important source of competitive advantage. Environmental regulations in Germany and Sweden led them to anticipate international consumer demand in car manufacturing and were a source of competitive advantage.

Porter recognizes that the state may fail, but he sees failures as due to the fact that government does not understand and address the key determinants of competitive advantage. Recently, some criticisms of Porter's treatment of the role of government have been advanced. This treatment neglects the political economy of public decisions and regulations. O'Shaughnessy (1996) argued that Porter

> tends to ignore the nature and force of the many political imperatives that lead to a deviation away from state spending on key 'factors' that might have aided in the creation of competitive advantage. . . . Governments, pressured as they so often are by lobby groups, sometimes seems incapable of making the strategic choices necessary to support and sustain the competitive advantage.

In this sense, his conceptual framework is incomplete and could be further developed by incorporating the analytical tools provided by public choice analysis. This view seems to be supported by the Italian case of retailing regulation.

6.4   Final remarks and conclusion

This paper analyses the determinants of international performance in the Italian pasta industry by using the appropriate conceptual framework provided by Porter's diamond. This industry is a success story. All trade indicators confirm the excellent export performance. The industry's export propensity and net exports are the highest of all the Italian food industries. However, despite its success in terms of exports, the industry is characterized by some points of weakness. The international performance of the industry is not dissimilar to that of other Italian food industries with regard to the slowness in the process of internationalization, the small number of firms with FDIs and the low degree of multinationality of firms engaged in this activities.

As we have seen, Porter emphasizes that advantages throughout the diamond are necessary for achieving a sustainable competitive advantage and excellence in international performance. The use of his approach produces results quite capable of explaining both the success in exports and the weakness in FDI of the Italian pasta industry.

Our analysis shows that while some factors have indeed worked positively and contributed to the positive performances in terms of exports, the Italian pasta industry could not count on the interaction of a larger number of factors and therefore on the mutually reinforced mechanisms which are at work in these cases. The sophistication of Italian final consumers and the exogenous shift in foreign demand (as a consequence of the diffusion of the Mediterranean diet and the competitiveness of domestic suppliers of pasta machinery) were the main determinants of export performance.

However, other crucial factors, such as domestic rivalry and the presence of advanced and sophisticated retailers, have played only a minor role. In both cases, this was the consequence of the fragmented structure of Italian grocery retailing. Our analysis confirms that the fragmented and backward Italian retailing sector has negatively influenced the dynamics and the evolution of the Italian pasta industry. This negative influence might have contributed to retard the industry's process of internationalization.

The positive role of retailing sector for dynamic efficiency is well recognized by Porter. Large powerful chains not only bargain away profits from manufacturers, but also stimulate them to cut costs, develop customer services and intensify product innovation.

A competitive environment characterized by the presence of large retailers and a concentrated retailing sector may contribute positively to competitive advantage, while its absence has a negative influence on the upgrading of competitive advantage, since a nation's firms are forced to learn how to deal with advanced and sophisticated channels abroad instead of at home. If proximity does matter, this absence is a factor of weakness.

We find that Porter's approach provides an appropriate conceptual framework to detail and explain the sources of competitive advantages and competitive performance of the Italian pasta industry. Our analysis, however, suggests the opportunity of putting more emphasis on some aspects which seem rather neglected in Porter's framework. First, there is a need for a deeper and more detailed analysis of the mechanisms by which retailing may exert its influence on market structure and firm's strategies. Demand conditions are not the only channel through which retailers may exert their influence. The above analysis indicates that powerful retailers can have a crucial impact even on other parts of the diamond such as the intensity of domestic rivalry. Both factors were at work in the case of the Italian pasta industry. This puts greater emphasis on the role played by retailers and on the negative consequences of the laggard nature of vertical relationships between Italian pasta manufacturers and retailers on the process of the industry's internationalization.

References

Bond, E.W. (1984) International trade with uncertain product quality. *Southern Economic Journal*, 51, July, 196–207.

Braga, F. and Raffaelli, R. (1995) Implications of a changing commodity quality definition: the case of Canadian durum wheat exports to Italy. *Agribusiness*, 11(5), 463–72.

Databank (1995) Report on pasta secca. *Competitor series B 39*, July.

Dunning, J.H. (1993) Internationalising Porter's diamond, *Management International Review*, special issue, 33(2), 7–15.

Eurofood (1994) Emerging markets attract over 50% of cross border M&A. *Eurofood*, September, 3–4.

Eurofood (1995) Italian consumer test shows good quality is possible at low prices. *Eurofood*, June, 5.

Eurofood (1996a) Barilla denies take-over rumours. *Eurofood*, 31 January, 8.

Eurofood (1996b) Boom in UK pasta market fuelled by fresh pasta. *Eurofood*, 29 August, 4.

Galizzi, G. and Venturini, L. (1996) Product innovation in the food industry: nature, characteristics and determinants, in *Economics of Innovation: The case of food industry* (eds. G. Galizzi and L. Venturini), Physica-Verlag, Stuttgart, pp. 133–53.

Kravis, I.B. and Lipsey, R.E.(1992) Sources of competitiveness of the United States and its multinational firms. *Review of Economics and Statistics*, 74(2), 193–201.

Largo Consumo (several issues) monthly publication, Editoriale Largo consumo, Milan.

Market Research Europe (1994) European markets: food, pasta. *Market Research Europe* March, 1–22.

Mediobanca (1994) *Richerche e Studi, R&S*, Vol. 1. Mediobanca, Milan.

OECD (1994) *Economic Studies*, 23, Winter.

O'Shaughnessy, N.J. (1996) Michael Porter's competitive advantage revisited. *Management Decision*, 34(6).

Pieri, R. and Venturini, L. (1995) *Strategie e competitività nel sistema agro-alimentare – Il caso Italiano*. Collana SMEA, Franco Angeli, Bologna.

Porter, M.E. (1980) *Competitive Strategy: Techniques for Analyzing Industries and Competitors*. Free Press, New York.

Porter M.E.(1990) *The Competitive Advantage of Nations*. Free Press, New York.

Quarleri, G. (1994) Pasta, anno primo dell'era discount. *Largo Consumo*, 9, 84–93.

Rolfo, S., Vaglio, P., Vitali, G. (1993) Small firms and technological development in the food processing machinery industry: the case of pasta and other cereal derivative machines in Italy. *Small Business Economics*, 5, 307–17.

Rugman, A.M. and Verbeke, A. (1993) Foreign subsidiaries and multinational strategic management: an extension and correction of Porter's single diamond framework. *Management International Review*, special issue, 33(2), 71–84.

Torazza, V. (1990) L'evoluzione della domanda di pasta di semola. *Largo Consumo*, suppl. 5, 44–8.

Traill, B. and Gomes da Silva, J. (1994) Measuring international competitiveness: the case of the food industry. *International Business Review*, 5(2), 151–66.

UNIPI (1992) *Annuario Generale dell Industria della Pastificazione in Italia, 1991*, UNIPI, Rome, May.

Venturini, L. (1994) *Endogenous sunk costs and structural changes in the Italian food industry*. Discussion Paper No 2, Structural Change in the European Food Industries, University of Reading.

7 Ecological minded retailers: a driving force for upgrading competitiveness of the Swedish food sector?

MAGNUS LAGNEVIK AND HELÉNE TJÄRNEMO

7.1 Introduction

The main question in this chapter is whether the ecological responsiveness of Swedish food retailers, together with some of the world's toughest food safety and animal husbandry legislation, is a possible driving force in raising standards and upgrading the competitiveness of the Swedish food sector. This study differs from the others in this book in using the Porter (1990) framework in order to forecast future competitiveness, and not to analyse a competitive industry or sector as it exists today. Thus the purpose is to explore, within a Porter framework, if the ecological responsiveness[1] of Swedish retailers is a driving force in the upgrading of competitiveness, in such a way that the creation of competitive advantage for the Swedish food sector is possible in the future. More specifically, we will analyse how retail strategies interact with other factors in the dynamic process that creates competitiveness.

In the analysis we use the 1991 Agriculture Canada definition:

> a competitive industry is one that possesses the sustained ability to profitably gain and maintain market share in domestic and/or foreign markets.

Thus we analyse the dynamic process that has created rapid growth of market shares for ecological food in the domestic Swedish market. We discuss how sustainable this growth is and to what extent it could lead to international competitiveness. Finally, we discuss if resources and competencies created in the ecological food sector could be valuable in the upgrading of competitiveness in other parts of the Swedish food sector.

7.1.1 Definition of ecological products

In a situation where words such as organic, biodynamic, biological, natural, ecological, organic-biological, environmental and sustainable are used more or less synonymously and the concepts of integrated production (IP) and

---

1. By ecological responsiveness we mean attitudes towards ecological issues as well as how ecological issues are incorporated into retail behaviour in, for example, planning and implementation of strategies and policies.

low-input sustainable agriculture (LISA) have been referred to as new sustainable alternatives to conventional agriculture, the need for a clear distinction between different production systems is essential (Lampkin, 1994).

According to Lampkin (1994) it is possible to distinguish between at least three different approaches to agriculture: conventional, low-input and ecological.[2]

Conventional agriculture is intensive and highly specialized farming which allows the use of chemical fertilizers and pesticides. Increased awareness of the negative effects of conventional agriculture on the ecosystem has opened up market prospects for more sustainable alternatives, i.e. production systems with either reduced use of chemical fertilizers and pesticides or ecological systems that rely on ecosystem management rather than external chemical inputs.

Both **low-input** and **ecological production systems** can be regarded as more sustainable alternatives. The major factor distinguishing between low-input and ecological production systems is that the former allows reduced use of chemical fertilizers and pesticides, while in ecological production systems these kinds of external input are eliminated. Ecological farming thus is an approach to agriculture that relies on ecosystem management rather than external inputs of chemical fertilizers or pesticides (Lampkin, 1994).

Low-input production systems, such as IP in Europe and LISA in the US, allow for conventional systems to be modified, ecological production systems entail a complete change in farming approach. The low-input approach can therefore be seen as a compromise between conventional and ecological production. A further distinction between the two alternative production systems lies in the existence of control systems, or as Lampkin (1994) writes:

> The major factor which distinguishes organic farming from other approaches to sustainable agriculture is the existence of both legislated and voluntary standards and certification procedures to define a clear dividing line between organic and other farming systems, primarily for marketing purposes. (p. 5)

Tate (1994) goes even further:

> A key element of the organic produce market is regulation. It is necessary in order to maintain the high ethical standards of the organic movement, to retain consumer confidence in produce, to encourage and support genuine organic farmers, and lastly to provide a basis for traffic in organic produce across frontiers. The symbol schemes that farmers, wholesalers and processors join are a central feature of organic regulation. Once certified by the association running the scheme, they are entitled to use the symbol. (p. 15)

From 1 January 1993 all fresh and processed plant products sold in the EU as 'ecological' (or 'organic/biological' equivalent) must meet the stan-

---

2. The term 'ecological' is used in the Nordic languages, German and Spanish; 'organic' is more commonly used in English, and 'biological' more often in, for example, Greek, French, Italian, Dutch and Portuguese (EC regulation 2092/91).

dards defined in EC regulation No. 2092/91 on ecological production of agricultural products and foodstuffs. The regulation does not yet include produce of animal origin (Tate, 1994).

The EC legislation emphasizes the role of the inspection system and each member state must have a designated inspection authority which implements and polices the inspection scheme alone or in conjunction with approved ecological certifying organizations (EC, 1991). In Sweden there are two ecological certifying organizations which are accredited to the Swedish Board of Agriculture: Kontrollföreningen för Ekologisk Odling (KRAV) and the Swedish Demeter. These two organizations are responsible for the control and certification of ecological farmers and products within Sweden. The purpose of the inspection system thus is to ensure the credibility of ecological produce and to serve as a guarantee of ecological products throughout the whole food chain from farmer to consumers. The inspection system thus is primarily for marketing purposes.

In this chapter we follow the KRAV definition (KRAV kontrollen, 1995) and include as 'ecological' products that are certified by KRAV or Demeter and carry the KRAV or Demeter label (see Figure 7.1). Since Sweden is a relatively new member of the EU, the rules of KRAV and Swedish Demeter still differ somewhat from the EC standard. One major difference is that KRAV also includes ecological livestock farming, and that ethical aspects of animal welfare, in addition to mere health aspects, are regulated: paying full regard to animals' behavioural needs, for example (KRAV regler, 1996)[3]. In Sweden KRAV has come to be used as more or less synonymous with ecological, whereas Demeter refers to the biodynamic production approach.

Figure 7.1 The KRAV and Demeter symbols. Reproduced with permission.

3. If farms that are connected to Demeter for crop or horticulture production also have ecological livestock, this part of the production is controlled by KRAV.

7.2 Factor conditions

In this section we describe the inputs to retail; ecological products, their distribution system and the specialized production factors of importance.

7.2.1 Development of ecological production and distribution systems

During the postwar period, ecological production has been carried out in parallel with modern conventional agriculture. However, it was not until the 1980s that ecological issue became a major societal issue and consumers in general began to pay attention, not only to health aspects of food products, but also to the way the food was produced, such as ecological and ethical aspects (Terrvik, 1995; Swedish Board of Agriculture, 1996).

To begin with, the supply of ecological products was small and scattered: there were large distances between growers and no grouping of the supply, and most products were sold directly from farmers to consumers. At the end of the 1980s there were the first signs of changing market conditions for ecological products. Swedish consumers wanted to be able to buy ecological vegetables in their usual grocery stores. The change from selling ecological products directly from farmers to consumers to selling through retailers brought about a major distribution problem. On the one hand, the supply of ecological products was still scattered among many, very small growers, and these individual growers were not in a position to establish contacts with major retailer chains. On the other hand the retailer chains were not used to handling many small suppliers. They preferred to have a few suppliers with large enough volumes and continuity of production in order to supply their large-scale distribution systems. This resulted in a paradoxical situation where ecological growers were not able to meet a ready demand while at the same time retailer chains could not find supplies of ecological products on a sufficient scale. This 'misfit' between demand and supply resulted in the establishment of a whole new distribution system for ecological products.

At the beginning of the 1980s there were two major organizations to which most ecological farmers or growers were connected, the biodynamic and the organic-biological (Sahlström, 1983). In order to group the different organizations that worked with ecological production, Samarbetsgruppen för Alternativ Odling (SAO) was founded on the initiative of Svenska biodynamiska föreningen, Förbundet organisk-biologisk odling and Friends of the Earth Sweden (SAO and Bokskogen, 1983). The purpose was to encourage alternative agriculture in Sweden, in the Nordic countries and internationally. SAO also tried to influence authorities and institutions that worked with agricultural questions. In 1985 Alternativodlarnas Förening (ARF) was founded, an organization of which all professional growers with an interest in ecological agriculture could become members. Besides gradually taking over the tasks of SAO, ARF was one of the initiators of the establishment of

KRAV in 1985. Today ARF (which changed its name in 1994 to Ekologiska Lantbrukarna i Sverige, the Swedish Ecological Farmers Association) works, among other things, in favour of an ecological orientation of agricultural policies, research and education, as well as influencing the rules of KRAV. Its work is financed by membership fees as well as by governmental subsidies. Neither SAO nor ARF however, worked directly with the establishment of distribution and marketing systems for ecological products.

As the demand increased for ecological products from actors within the large-scale distribution system, wholesalers and retailers, the need to organize the supply of products, not only the ecological farmers, become more and more necessary.

Three different producer co-operatives were established with the purpose of developing distribution and marketing channels for ecological produce. Samodlarna, already established in 1983, co-ordinates and markets organic fruit, vegetables and potatoes. In 1992 Eco Trade was founded for grain and oil plants and in 1993 Ekokött started to co-ordinate and develop marketing channels for ecological meat. Thus, a specialized marketing and distribution system for ecological food products has developed (see Figure 7.2).

The main purpose of Samodlarna, Eco Trade and Ekokött is to co-ordinate and market ecological products better. All three organizations contract and act as intermediaries within their respective product group to the food industry, wholesalers and retailers. The organizations work only with KRAV-certified farmers and products, and Ekokött utilizes only KRAV-certified slaughterhouses and processing companies.

The specialized marketing and distribution system handles a large share of the ecological produce, but not all of it. For example, ecological milk products are sold and distributed by the conventional dairies (both under their own brands and as retailer private brands) directly to the stores and the food manufacturers as a complement to conventional milk products. Retailers and food manufacturers may also have their own contracts with ecological farmers, for example for vegetables. This produce is not co-ordinated by the specialized marketing and distribution system.

Besides the organizations working directly with the distribution of ecological products there are a number of different non-profit associations that work with information and political lobbying, such as Ekologiska Lantbrukarna i Sverige, Sveriges Ekomjölks Bönder (SEMB), Biodynamisk Förening, Rådgivarföreningen Ekologiskt Lantbruk and the 10 % Campaign.

7.2.2 The growth of ecological production

The traditional way of defining market share is to consider the value or amount of products sold on a market. In this chapter we define market share in quite a different way, as a share of ecological cultivated area or production. The main reason for this definition is that the figures for value and

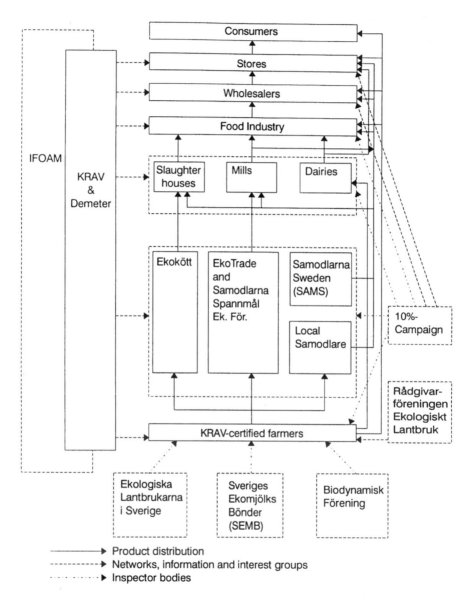

Figure 7.2 A brief illustration of the structure of the ecological production system. (Not included: catering and export.)

amount of ecological products sold are hard to obtain. (In those cases where it is possible to obtain the share of total sales, and not production, these figure will be referred to specifically.)

The area of ecological agricultural land[4] grew slowly between 1985 and 1988, but in 1989 there was an immense increase. What influenced this increase is hard to say for sure. At least two possible factors ought to be mentioned: the growing interest in ecological products from a retailer perspective and a new government subsidy for ecological production which was granted for the first time in 1989. During the following years the share of ecological agricultural land increased from approximately 0.5% in 1989 to 3.6% in 1995 (Figure 7.3) (KRAV kontrollen, 1995). By the end of 1996, ecological agricultural land was expected to reach a share of almost 5% of the total agricultural land in Sweden (Gardfjell, 1996).

The area of ecological cultivation is estimated to continue growing between 1996 and 1999. According to the Swedish Board of Agriculture (1996) it is not certain that there will be a corresponding increase of ecolog-

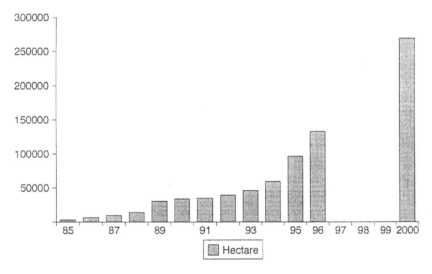

Figure 7.3 Ecological agricultural land in Sweden (hectares) (from KRAV kontrollen for 1995, estimation for 1996 and governmental goal for the year 2000.

4. When we refer to 'ecological agricultural land' we follow the KRAV definition which includes agricultural land that is (a) certified by KRAV (about 47% of total ecological agricultural land); (b) agricultural land that is in conversion from conventional or low-input production systems to ecological production and controlled by KRAV (about 35%); (c) agricultural land that is connected to both KRAV and the Swedish Demeter (about 4%); (d) agricultural land that is connected only to Swedish Demeter (less than 2%); and agricultural land that is supported by governmental eco-subsidies but not controlled by KRAV or Demeter (about 12%). (The control of this production is done by the county administrative board which also administers the subsidies.) The ecological agricultural land which might not be included within this definition is primarily small acreages (too small to be supported by governmental subsidy) for private domestic purposes. The main reason for this is the lack of statistics of ecological production apart from the inspection system.

ical products on the market, since some subsidy-takers do not produce for the ecological market and thus are not certified by KRAV or Demeter.[5] However, as this share of agricultural land is only about 12% (see footnote 4) we do not think it will be a limiting factor for the development of ecological products.

For most crops the share of ecological production is still only a few percent of total production of a specific product. This is the case for grain (1.1%), pasture (2.6%), potatoes (2.8%) and sugar beet (3.5%). The share of KRAV-certified vegetable cultivation is, however, more than 12% (KRAV kontrollen, 1995). According to the Swedish Board of Agriculture (1996), the high share of ecological vegetable production was predominantly in outdoor cultivation. Even though the share of ecological cultivation is still very low the share is growing. Table 7.1 indicates the growth of KRAV-certified production from 1993 to 1995.

In 1995 the demand for ecological grain greatly exceeded the supply. In order to let the market continue to grow, Eco Trade began importing thousands of tonnes of ecological grain. As a result of indicated higher prices of ecological grain most industry customers decided to cut their quantities and many new market actors chose to refrain from launching new ecological products (Eco Trade, 1995). Even though the need for imports was reduced, the lack of national supply led to the fact that every third delivery of grain via Eco Trade was imported. In 1996 there was a better fit between demand and supply, at least for some crops. The lack of supply is especially great for high protein wheat suitable for bakeries (Eco Trade, 1995).

Larger supply may, however, lead to lower prices for the farmer. Thus Eco Trade is in a dilemma. On the one hand, they need a large supply in order to

Table 7.1 Share of KRAV- certified production compared with Swedish total 1993–1994 (Figures for vegetables are from 1990 and 1993)

	1993	1994
Grain	0.89	1.1
Pasture land	1.98	2.6
Potatoes	2.0	2.8
Vegetables	10.00	12.3
Sugar beet	0.13	0.3
Others	3.25	3.5
Total	1.3	2.0

Source: KRAV kontrollen, 1994 and 1995.

5. Ecological produce that is not grown for the purpose of marketing it as ecological, for example, produce for one's own domestic use (on the farm or in a household) or pasture land, does not need to be inspected and certified by KRAV or Demeter. This produce is, however, nonetheless ecological if it is in accordance with the general definition of ecological production as regulated in the EC regulation.

encourage the food industry to develop ecological products. On the other hand, they do not want to have surplus grain because then the price to the farmer becomes much lower, which does not encourage the farmer to produce ecologically. There are also threats from outside Sweden. For example, there are surpluses of various crops in, for example, Denmark and Germany. Thus, if the industry customers find the prices too high in Sweden they may buy directly from abroad.

In 1988 the first KRAV rules regarding ecological livestock farming and meat production came into force. During that year about 25 livestock farmers were certified. There was, however, one major problem. According to the rules of KRAV the ecologically reared animals had to be slaughtered at KRAV-certified slaughterhouses, of which only two or three existed in 1988. The majority of the ecological meat thus had to be sold as conventional (Holm and Drake, 1989). Today, however, there are KRAV-certified slaughterhouses in many parts of Sweden.

Demand for ecological milk and meat has given rise to increased ecological livestock farming and the number of KRAV-certified livestock farmers has doubled every year since 1993 (Figure 7.4).

It is mostly cows, calves and cattle, lamb and poultry that are controlled by KRAV (see Table 7.2). Rearing of KRAV-certified pigs is still limited. This is mainly due to the special requirements of outdoor pasture, different forage and longer suckling period, differing in many respects from conventional pig rearing. It requires large areas of pasture land for grazing, reconstruction of stables and ecological feeding stuffs (Swedish Board of Agriculture, 1996).

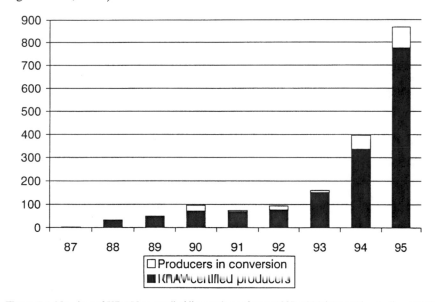

Figure 7.4 Number of KRAV-controlled livestock producers 1987–1995 (KRAV kontrollen, 1995).

Table 7.2 Number of KRAV-controlled animals 1994 and 1995

	1994	1995
Cattle	2397	6151
Dairy cows	4237	8983
Calves and young cattle	8399	21 992
Sows	321	559
Pigs, piglets	2977	6502
Sheep	4253	8567
Lamb	6673	11 560
Laying hens	n.a.	10 541
Others	n.a.	2242

Source: KRAV kontrollen, 1994 and 1995.

Figure 7.5 shows that the numbers of ecologically reared animals sold to slaughterhouses have increased for almost all species. In spite of this increase, the total quantity of meat is not estimated to meet the demand, at least not in all parts of Sweden. In 1995 the proportion of ecologically reared animals slaughtered in Sweden was 0.05% pigs, 1.5% lamb, 1% calves and 0.5% cattle (Ekokött, 1996).

During the last few years there has been an increase in KRAV-certified dairies in Sweden, and as a result the market shares of ecological milk and yoghurt products have increased from 1.0% of total volume sold in January–March 1995 to 1.4% in January–July 1996 (Thulin, 1996). The production of other ecological dairy products, such as cheese, is still in its infancy.

The development of ecological production has moved from basic primary products, such as fruit and vegetables, to processed products, such as flour and milk, and finally to even more refined products such as cheese and baby food. One explanation for this development is that ecological primary products are a prerequisite for more refined products. It is impossible to produce ecological milk if there is no supply of ecological animal feed, and in order to make ecological cheesecake the manufacturer needs ecological eggs and flour. Thus, a wide assortment of ecological primary products is needed in order to produce more highly processed food products. Higher value ecological products are also less price sensitive than primary products, since the share of the ecological primary product in the total consumer price of the refined product is smaller. For example, a higher price for ecological potatoes has a limited impact when it comes to producing ecological snacks. Today the assortment of ecological products is wide, including not only basic primary products but also processed products such as ice cream, ready-to-eat meatballs, baby food and fruit-syrup.

In Sweden the average size of the organic farm is larger than that of the conventional farm, 51 ha compared to 31 ha (KRAV kontrollen, 1995). This implies that not only small farmers or part time farmers are involved with ecological production.

7.2.3 Conclusions

In order to achieve a working market for ecological products in Sweden the development of specialized distribution and marketing channels has been important. These intermediaries have gathered a scattered supply of ecological produce and made large-volume distribution possible. Samodlarna, Eco Trade and Ekokött have also agreed upon the rules and definitions of KRAV as a standard for ecological products which makes it easier to communicate not only with farmers, but also with the food industry and food retailers. Despite the fact that the share of ecological production continues to grow, there are still big differences in the continuity of supply between different products. For some products, especially perishables, periods of lack of

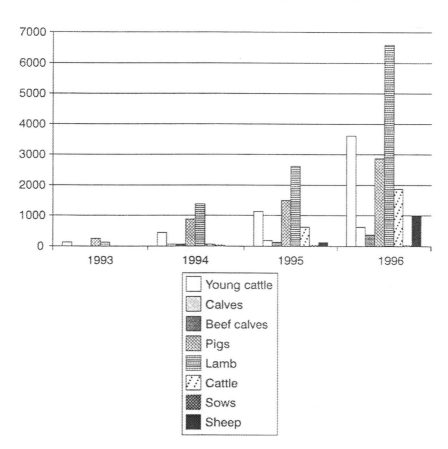

Figure 7.5 Slaughtered quantities of KRAV-certified animals (number of animals) (Ekokött).

supply make it impossible for food retailers to offer these products on a regular basis. Lack of supply of ecological primary products also makes it difficult to produce more highly processed ecological products. For other products there is sufficient supply or even, from time to time, supply that exceeds demand. Surpluses make it difficult for producer co-operatives to encourage farmers to convert to ecological production since farmers do not know if they will get a premium price for their produce, or in the worse case, be able to sell the produce as ecological at all.

It is probably impossible to balance demand or supply at every time and stage of development. The interesting question is whether the imbalances will finally result in a virtuous circle, where periods of lack of supply encourage more farmers to convert and are followed by periods of surplus which encourage the food industry and retailers to develop new ecological processed products.

Since ecological production needs to be inspected by KRAV through the whole food chain from farmers to food stores, a limiting factor has sometimes been lack of KRAV-certified intermediaries, such as mills, dairies and slaughterhouses. Stores that want to package and/or sell ecological products, such as fruit, vegetables, meat and cured meat by the piece also have to be certified by KRAV. The number of KRAV-certified intermediaries and stores has, however, increased during the last few years (see Figure 7.6), which ought in future to reduce the limiting effects of distribution.

Recently, the first Swedish KRAV-certified restaurant was opened, meaning that the restaurant offers at least one dish per day that includes KRAV-certified ingredients.

The specialized distribution and marketing system that has been developed for ecological products is a major factor positively influencing the market potential for ecological products, while the occasional limited supply of primary ecological products as well as lack of KRAV-certified processing companies may be factors influencing the market potential in a negative way. Another risk factor, if there are long periods of surplus for some primary products, is that fewer farmers will convert, or ecological farmers may even reconvert, due to lower price premiums. There appears to be a continuing driving force within the food chain for ecological products. It is important that the different actors develop at the same pace, or at least follow each other closely in time, and also co-operate with the actors down- and upstream in the food chain.

7.3 Demand conditions

There is a lack of studies on consumers' attitudes and behaviour regarding ecological agricultural products in Sweden. One reason for this is that most studies on ecological production and products have been carried out by stu-

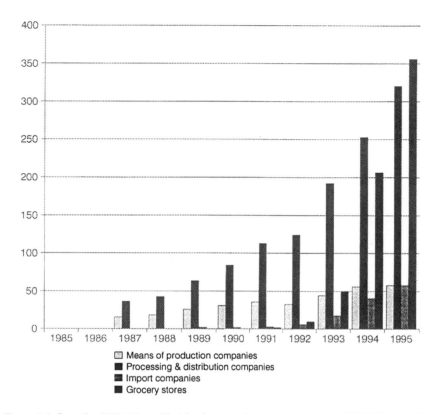

Figure 7.6 Growth of KRAV-certified food companies and grocery stores (KRAV kontrollen, 1995).

dents or researchers at the Swedish agricultural schools, thus focusing more on the farmer and production levels than on consumers. The few consumer studies that have been done have been concerned with specific product groups such as meat or vegetables. Some conclusions that have been drawn will, however, be presented in this section.

Mathisson and Schollin (1995) concluded that the recognition of the KRAV symbol has increased throughout the 1990s (Figure 7.7). In 1989 only 20% of Swedish consumers recognized the KRAV symbol while at the end of 1994 the figure was about 75%.

Furthermore, Mathisson and Schollin (1994) concluded that consumers' confidence in ecological products, as well as consumers' recognition of the KRAV symbol, was related to the general environmental profile of the grocery store of which they were customers. The higher the environmental profile of the store, the greater consumer confidence in ecological products and recognition of the KRAV symbol.

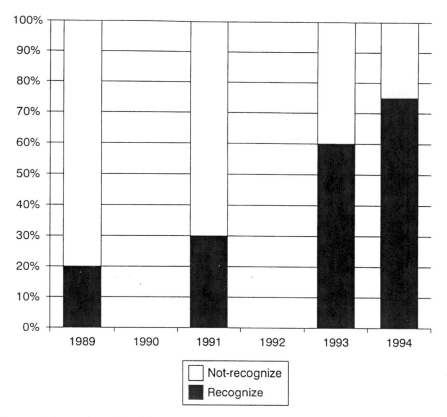

Figure 7.7 Proportion of Swedish consumers recognizing and not recognizing the KRAV label, 1989–1994 (Mathisson and Schollin, 1995).

Even though most Swedish consumers recognized the KRAV symbol, only about 40% of consumers bought ecological products more or less regularly (Figure 7.8). About 12% were really 'devoted' to ecological products and bought them often. The most significant fact about the 'devoted' consumers was that they were less than 35 years old. In addition to the 'devoted', 19% of consumers bought ecological products sporadically, and 9% bought ecological vegetables and potatoes regularly (but no ecological products within other product groups). The remaining 60% of Swedish consumers were either unaware (19%) of the fact that they sometimes bought ecological products (mostly men under 35 years of age who did not recognize the KRAV label) or consumers who almost never bought any ecological products (41%) (LUI/Samodlarna Sverige, 1994).

In a study of consumer attitudes towards organic meat production (Holm and Drake, 1989) a majority of Swedish consumers were highly concerned with the living conditions of animals (see Figure 7.9).

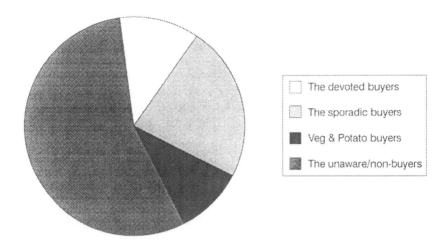

Figure 7.8 Four clusters of consumers with regard to their buying of organic products. (LUI/Samodlarna Sverige, 1994).

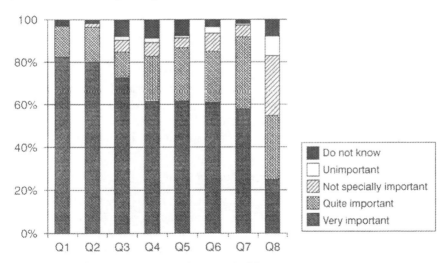

Figure 7.9 Swedish consumers' attitudes towards different ethical aspects of livestock living conditions (after data from Holm and Drake, 1989). Q1, low levels of sick animals; Q2, animals should not be contained or placed in small cages, but given a chance to develop a natural behaviour pattern; Q3, genetic technology and hormones should not be used; Q4, animals should be slaughtered in a more considerate way; Q5, preventive and routine use of medicine should not be allowed; Q6, animal fodder should be produced without pesticides and chemical fertilizers; Q7, animals should be able to be outdoors the greater part of the year; Q8, animals should live their whole lives on the same farm and receive fodder from their own farm.

7.3.1 Conclusions

The demand from consumers is substantial and growing. For many types of products demand exceeds supply for the moment. Forty per cent of the consumers buy ecological products on a more or less regular basis and 12% can be classified as devoted customers. This is a large segment of the food market. We also know that the devoted customers are usually under 35 years of age. Unless the ecological trend is reversed for some reason, the percentage of devoted ecological consumers will increase as the population grows older.[6]

If Swedish consumers are in the vanguard of ecological consumption, this could be an early indication of a market segment that will in time be developed in other European countries. If this is the case, we are talking about substantial numbers of consumers.

7.4   Firm strategy, structure and rivalry: a retailer chain perspective

In their role as intermediaries between manufacturers and consumers, food retailers are extremely important for the development of ecological products. Retailers have the opportunity to influence consumers by in-store and out-of-store marketing activities. They also have the possibility to persuade or direct manufacturers to produce more environmentally friendly products. In Sweden about two-thirds of all ecological products are sold through traditional convenience stores and supermarkets (Hagman, 1995). One reason for this large share is the introduction of their own ecological brands by the major retail chains. Another reason may be the small number of specialized natural or health food stores in Sweden. In spite of a very small market share, ecological products are sold through most of the major retailer chains in Sweden.[7] The major growth, however, began when the three major retailer groups introduced their own ecological brands in the beginning and mid 1990s.

7.4.1   Highly concentrated Swedish food retailing

Swedish food retailing is highly concentrated into a few major retailer groups; the merchant owned ICA, the consumer owned Konsum, the privately/merchant owned D-gruppen, the privately owned Axel Johnson, and the local, privately owned Bergendahl & Son. Wholesaling is highly integrated: ICA Handlarna supplies only the stores within the ICA retailer group while KD supplies only co-op stores. Dagab and Bergendahls can the-

6. It has been argued that the environmentally conscious consumers will change their consumption patterns to a more traditional profile when they grow older. In our view, however, there are more indicators showing that environmental awareness is a growing phenomenon. Young consumers have grown up in another environment and in another public debate. It is more reasonable to argue that they will base their judgement in the future on their own first-hand experience rather that that of their parents. Also, recent studies among Swedish food industry executives reveals that they expect environmental issues to influence food consumption, sales and production more and more (Wennerholm et al., 1996).
7. This is true for ecological products that are certified by KRAV. Biodynamic products, i.e. products with the Demeter label, are sold mostly through natural and health food stores, and only a few stores within the major retailer chains offer these products.

oretically sell to all stores regardless of retailer group, but in fact neither ICA nor co-op ever buys from these wholesalers. Within each retailer group there is a big variety of store types, such as supermarkets, discount stores and traditional small and medium-sized grocery stores (Figure 7.10). Together these five retailer groups control more than 75% of the Swedish food retailing market, both in value of sales and share of food stores. The remaining 25% of food is sold through, for example, petrol stations and bakeries (Svensk Dagligvaruhandel, 1994–1995).

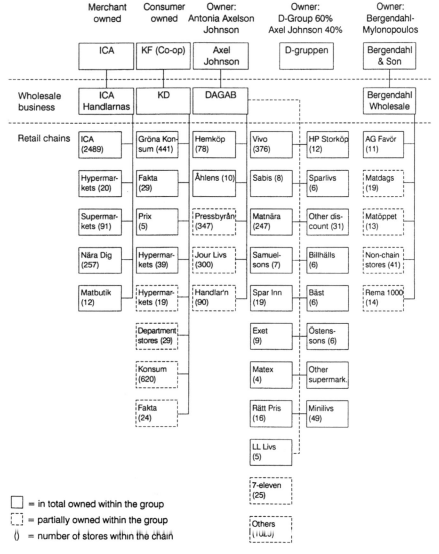

Figure 7.10 Swedish retailer sector 1994–1995. Data from Svensk Dagligvaruhandel. Fri köpenskap. 1994–1995.

It is an observable fact that many retailers have made the ecological issue a central plank in their marketing and competitive strategies. Besides offering ecological private brands, they have invested in freon-free refrigerators, trained their employees in environmental issues, supported projects that aim to protect lakes and forests from environmental threats, and much more. This 'greenness' can be seen as a way of changing the role of retailers from a passive role of logistics and retailing to a more active role in satisfying consumers' needs, as well as moving the firms from a competitive position of 'price wars' to differentiation. By positioning themselves as environmentally friendly, many retailers try to improve their images, achieve consumer loyalty and gain competitive advantage.

In Sweden at least three retailer chains (Gröna Konsum, Hemköp and ICA), have reoriented towards more environmentally apt strategies and behaviour during the last few years. In the next sections we will describe how the key actors in the highly concentrated Swedish retailing sector have all developed their strategies in such a way that ecological considerations play a very important role.

7.4.2   The eco-strategy of KF and Gröna Konsum

The increased environmental awareness among consumers together with a desire among KF's managers to improve the image of the retailer group resulted in a decision in 1990 to formulate an environmental policy. The ambition level was high: KF was to become the most environmentally friendly retailer group in Sweden. The first decision was to introduce at least one environmentally friendly product in each product group, this decision was hard to implement owing to lack of products. Furthermore, it was difficult to persuade existing manufacturers to develop environmentally friendly alternatives. KF therefore found it easier to develop its own ecological brand, and in 1991 the Änglamark brand was introduced, including 10 food products and 10 non-food products (for example chemical–technical products). One prerequisite was that ecological food products that were to be branded as Änglamark should meet the standards of KRAV.

A major problem was that the supply of ecological products was scattered and the purchasing department of KF was not used to handling many small, local producers. Nor were they used to discussing and evaluating the environmental aspects of products. There was also a need for a much closer relationship with the suppliers of ecological products than with suppliers of conventional products (Terrvik, 1995).

Although many product and store managers were doubtful about the financial profitability of the product range, Änglamark has become a market success and the number of products as well as the sales has increased every year. At the end of 1995 Änglamark included more than 140 products. The overall sales growth was about 65% between 1994 and 1995, and for perishables it was 138%. More than 50 products were introduced in 1995 and at the beginning of 1996 a baby food product line was introduced (KF Verksamheten, 1995). The introduction of Änglamark also resulted in a great deal of media publicity as well as much support for the product line from the Swedish Society for Nature Conservation and KRAV (Terrvik, 1995).

The development of Änglamark was, however, only a part of the major environmental reorientation of KF during the 1990s. At the end of 1993 KF decided on eight points by which the environmental agenda should be carried through within KF (KF Miljörapport, 1994, translated by the authors):

1. The activities and operations of the consumer co-operative shall be developed in such a way that they are based on the principle of sustainability.
2. The consumer co-operative shall educate its employees in environmental issues and through information contribute towards an increased environmental awareness among members and consumers.
3. The products and services that are offered by the consumer co-operative shall meet high environmental standards.
4. Transportation and handling of goods shall be made as resource conserving as possible and by using means of transportation that have the least possible negative effects on the environment.
5. The administrative activities, buildings and stores of the consumer co-operative shall be as environmentally adjusted as possible.
6. The own production of the consumer co-operative shall be produced in accordance with high environmental standards.
7. The consumer co-operative shall take an active part in the environmental debate and work on behalf of that society through regulations and other means, to give companies and consumers the pre-requisite of working in accordance with the principle of sustainability.
8. The consumer co-operative shall in its international activities work for increased international co-operation in environmental issues in accordance with the basic ideas of the co-operative as well as the intentions derived from UN's environmental conference in Rio.

This policy is valid for KF as a whole, that is for the wholesale business as well as all the different retailer chains, but one of these chains, Gröna Konsum, is more oriented towards the environment than the others. Gröna Konsum includes more than 400 stores with the market's largest assortment of ecological products. In a 1996 project called 'Sila kamelerna',[8] Gröna Konsum mapped each activity of its retailing operations with regard to its effects on the environment. The purpose of the project was, among other

8. Look after the big issues and let go of the smaller ones.

things, to integrate the environmental issues in the organization as a whole and to increase awareness among the employees.

Within Gröna Konsum there are specific policies regarding the in-store marketing of ecological products. For example, ecological fruit and vegetables are placed on the same counter as conventional products. All ecological products are marked by signboards above the counter and by signposts, in the form of a cloverleaf, on the shelves. Only ecological products that are KRAV labelled are included in Gröna Konsum's ecological range.

Within the chain some stores are chosen to become what Gröna Konsum calls 'spearheads', i.e. leaders in environmental concern within the chain as a whole. These stores are used as test stores for new Änglamark products as well as having to meet higher environmental standards than the other stores. For example, the employees have been more highly trained in ecological matters related to the range, and the stores must always have ecological meat and bread in their range as well as a large variety of ecological vegetables.

7.4.3   The eco-strategy of ICA

In 1992 ICA introduced its own ecological brands, Sunda for food products (1993 for vegetables and fruit) and Skona for chemical–technical products. Today there are about 30 ecological food products within the Sunda brand as a whole. According to Björn Hacklou (1996), who is responsible for the Sunda brand, sales have doubled each year since its introduction. Furthermore, most Sunda products are produced in Sweden, only 5% being imported. Hacklou estimates that Sunda has a share of approximately 95% of the total ecological food assortment within ICA.

For ICA, as for KF, the development of an ecological brand is only part of a major environmental orientation. From ICA's environmental policy it appears that ecological aspects are considered not only in planning the range of products, but also with respect to the total operation of the wholesaling and retailing business. The ICA retailers adopted their first environmental policy in 1989. At the end of 1996 this policy was revised in conjunction with the gradual introduction of a clearer environmental management system within the organization. The objective of ICA's environmental policy is that

> The ICA store shall both be (and be perceived by its customers and interested parties in its local community as being) responsible and progressive in its efforts towards environmental preservation. ICA Handlamas AB shall, by actively supporting the retailers in all of their jointly operated activities, continuously improve the environmental situation in a commercial and economically defensible manner. Our company shall both be and be perceived as being, responsible and progressive in its efforts towards environmental preservation. ICA retailers and ICA Handlamas AB shall work together to achieve long term sustainable ecological and economic development. (ICA Handlamas AB, 1997)

Included in this handbook are the following statements.

1. We shall conduct our business in accordance with current legislation and achieve the environmental requirements stipulated by the authorities by a good margin.
2. ICA Handlamas AB shall formulate action and remedial programmes which are followed up by environmental management systems and regular environmental audits.
3. Our actions are based on the fundamental preconditions for sustainable development set by The Natural Step.
4. ICA retailers' initiatives in their local environment with regard to product range, technology/operation, and ecological awareness shall be clearly apparent to their customers.
5. We shall aggressively market what we are doing to achieve a better environment by targeting customers, moulders of public opinion, authorities and colleagues.
6. We shall co-operate with local authorities and organizations so that we formulate our decisions in accordance with the environmental objectives of society and using the optimum environmental know-how available.
7. We shall work to ensure that environmentally adapted alternatives are available within all product sectors.
8. We shall work to ensure the removal from our range of products which are dubious from an environmental standpoint.
9. ICA Handlamas AB shall, when necessary, develop and supply its own environmentally friendly products.
10. ICA Handlamas AB shall work towards promoting reduced packaging quantities and to prioritizing the use of environmentally adapted material which can be recycled.
11. We shall reduce the environmental impact from warehousing and distribution by reducing our use of power, chemicals, packaging materials etc., by sorting waste products at source and by improving and increasing the efficiency of transportation.
12. We shall take environmental considerations into account when purchasing raw materials and services for our day to day activities.
13. ICA Handlamas AB shall supply technology, skills and services, which enable an ecologically defensible and financially reasonable development in stores, subsidiaries and other aspects of our operations.
14. We shall provide information on and actively market environmentally adapted products.
15. We shall display our environmental efforts openly, both internally and externally, through clear communication.
16. We shall train and involve all employees in the environmental work on an ongoing basis.

7.4.4   The eco-strategy of Hemköp

Hemköp started its ecological reorientation in the early 1980s. After a debate in Sweden concerning husbandry practices, Hemköp decided to stop buying the meat of pigs that were put under stress. In 1989 Hemköp started to co-operate and contract with livestock farmers who were more concerned about the well-being of the animals than farmers in general, in order to control the meat that was sold in their stores. In 1993 they dissociated themselves from imported meat due to the wide prevalence of salmonella and the use of antibiotics in animal feed outside Sweden.

Hemköp is positioned towards quality, health and environment, and tries to communicate an environmentally friendly profile. It successively replaces environmentally dangerous products in favour of more environmentally friendly alternatives. It gives priority to environmentally friendly products in marketing activities and to local suppliers when purchasing. Environmentally friendly products are also sold at lower margins than conventional ones in order to stimulate buying. Hemköp is a supporter of several environmental projects. In Hemköp's environmental policy it is stated that:

> Hemköp is going to contribute to a sustainable development that is based on the theory of the eco-cycle. Environmental concern is to be considered in every decision at all levels in the organization. (Hemköp's environmental policy, 1995, translated by the authors)

The policy focuses on three different levels of the organization; the stores in their relations with the consumers, the stores internally, and the office internally.

For the stores in their relation with the consumers, the following guiding principles are mentioned:

1. Get rid of non-desirable products.
2. Try to influence the suppliers in order to get rid of unnecessary packaging materials.
3. New products in the range ought to meet with the demand for an eco-cycle society.
4. Continue with education and information to those environmentally responsible and other employees.
5. Establish local contacts with the Swedish Society for Nature Conservation and similar organizations.
6. Encourage the use of paper and cloth carrier-bags.
7. Give priority to KRAV-certified food products and local suppliers.
8. Sell as much as possible by the piece.
9. Arrange for delivery of cardboard boxes in the store.

For the stores internally the guiding principles are:

1. Continue with education and information to those environmentally responsible and other employees.

2. Sort the refuse.
3. Use low-energy bulbs.
4. Control refrigerators and freezers (freon).
5. No non-returnables in the lunch room.
6. Clean the store with environmentally friendly products.
7. Try to influence the suppliers in order to get rid of unnecessary packaging materials.
8. Think 'environment–health–quality'.
9. New products in the range ought to meet with the demand upon a eco-cycle society regarding display and pricing.

For the office internally the guiding principles are:

1. Sort paper.
2. Use low-energy bulbs.
3. No non-returnables in the kitchen.
4. Use environmentally friendly cleaning products.
5. Use environmentally friendly office supplies.

All Hemköp stores (about 80 in 1995) are KRAV certified.

During spring 1996 Dagab, the wholesaler supporting the Hemköp chain, introduced its own ecological private brand Fauna, including 10 products (food or chemical–technical products). Hemköp was one of the major chains which adopted parts of this product concept. By the end of 1996 the Fauna range included 15 products.

7.4.5 Conclusions

A key observation is that the major retailers in Sweden have adopted ecological strategies and made them an important and integrated part of their retailer image.

Since a large share of the distribution of ecological food goes through traditional retailers, this implies the possibility of selling large volumes to consumers. Another important aspect of the retailers' strategies is that they all have well developed marketing functions. Since ecological aspects, together with taste and price, have been key components in the marketing of food in newspapers and especially on television for a few years now, a number of well formulated ecological concepts have reached the consumers.

The retail chains have had success with their ecological brands. All retailers have, in addition to their own brands, also labelled the products with KRAV. This voluntary concept and control procedure has thus become an institutionalized certificate for ecological products. From the consumer's point of view, ecological means KRAV-certified, even if other environmentally friendly concepts are competing in the European arena.

For some of the retailers, the ecological brand has also been tied to Swedish agricultural products, which induces nationalistic feelings about ecological products. This process has been supported by a huge nationwide campaign from the Swedish farmers' association, where Swedish agricultural products were said to be the best in the world, or possibly on the way to becoming the best in the world, as far as environmental qualities are concerned.

7.5 Eco-strategy implementation in retail stores: problems and possibilities

In spite of the overall environmental policies and the ecological brands of the three major retailer groups, there are differences between different retailer groups and retailer chains as well as within the same retailer chain in the ecological orientation of stores. Store owners and store managers have a great deal of freedom to choose what products to buy from the wholesaler or their local farmers. In this respect much of the demand is a reflection of the attitudes and willingness to deal with ecological products at a store level. The store owners' and managers' attitudes and willingness to offer ecological products can depend on many factors, for example, the location of the store, the size of the store and the attitude towards the characteristics of ecological products. In this section we present some opinions regarding ecological issues from 21 store owners or managers within Gröna Konsum, Hemköp and ICA.

The interviews were conducted between January and March 1996 at the stores. The interviews lasted 1–2 hours and dealt with subjects such as attitudes towards ecological issues in general and ecological products specifically, the ways ecological considerations were integrated in strategy, marketing and daily operations, etc. All but two interviews were audiotaped and then transcribed. When taping was refused or not feasible, notes were taken and transcribed the same day.

The reason why we devoted so much interest to the opinions and experience of the store owners and store managers, is that we were interested in how ecological strategies can be implemented. Only if a strategy of the retail chain is meaningful for the store owner and store manager will it be successful in the long run. Ecological matters must play a role in the everyday life in the store and the strategy must generate money and sales that make it a worthwhile part of the corporate strategy. Furthermore, it is our hypothesis that the success of the ecological strategy is very much dependent on how the store owners and store managers lead, motivate, communicate and care for the ecological strategy. The detailed analysis of this aspect will appear in a forthcoming report, and here we merely illustrate some concrete pros and cons of the ecological strategy.

The attitudes and experiences of store owners and store managers towards ecological products and the implementation of eco-strategies is presented under four different headings covering product appearance and availability, marketing activities, internal commitment as well as external stakeholders (for example, consumers and authorities).

7.5.1 The role of product appearance and availability

The store owners and managers thought that the quality of ecological products has become better during the last few years. Some of those interviewed still felt that ecological fruit and vegetables sometimes had a less attractive appearance; for example, they may look rough or uneven. The store owners and managers did not think that ecological products tasted any different. They were satisfied with the quality of non-perishable products, especially for products within the retailers' own brands.

One factor that may explain the occasional less attractive appearance of ecological products is the limited supply. When the supply is limited there is no need to be critical and choose only high quality primary products; there is no need to be fastidious about the product's appearance, as long as it is produced ecologically.

Even though the supply of ecological produce in general has become larger there is still seasonal and/or regional lack of supply.

Another related problem mentioned by some of those interviewed was that they had to buy quantities of ecological meat which were too large for the size of their stores. In addition, they could not be selective in what cuts of meat they wanted, but had to buy the quantities and cuts that were offered to them by the meat industry. Two store managers had solved this problem by splitting one delivery between their stores.

One store manager found the ordering procedures for ecological products a problem. For ecological fruit and vegetables he had to order from the wholesaler 1–2 weeks in advance, whereas he could normally order conventionally produced fruit and vegetables the day before delivery. He found it difficult to estimate how much he would be able to sell such a long period ahead, and had to face the risk of either buying too much or too little. This may be one reason why some of those interviewed mentioned that they preferred to buy ecological vegetables directly from the farmers instead of through their wholesalers.

7.5.2 The role of marketing activities

Price. The consumer price of ecological products in general is higher than that of similar conventional products. Many of the store owners and managers interviewed thought the consumers were willing to pay a price premium of 20–25% on ecological products in general. When the price difference is larger, many of those interviewed found it hard to argue in favour of the more expensive ecological product in their contacts with consumers. In spite of some ecological products being much more expensive than their conventional counterparts, those interviewed wanted to offer their customers ecological alternatives. In some stores they had chosen to buy very small quantities, in order to reduce the risk of discarding the products and/or reducing the margins in order to get the products sold. One store manager claimed that he lost money on every KRAV-labelled banana he sold, but that he still wanted to keep ecological bananas in the product range.

In-store marketing activities.   In all three chains it is a policy to give priority to KRAV-labelled products when it comes to providing shelf-space. Many of the store owners and managers interviewed mentioned that they sometimes make the KRAV collection even more visible and gather the products all together in a special section or aisle end position. According to some of those interviewed, it is a policy in their stores to put ecological products both next to the leading brand within each product group and together in a special section.

External marketing communication.   The store owners and managers interviewed emphasized the importance of not having to position their stores towards price. Offering ecological products provided the stores with a new argument in their external communications with consumers, more feeling and less price. Some of those interviewed told customers in their advertising about the environmental considerations and actions they had taken since last time. One store manager pointed out that he needed to remind customers that the store was environmentally concerned.

The store owners and managers considered it important to receive publicity for environmental activities implemented and performed within the store. One of those interviewed felt that publicity perhaps did not result in higher sales of ecological products *per se*, but that it resulted in a positive image of the store in general and as such was important for the store.

7.5.3   The role of internal commitment

Implementing an eco-strategy may be considered demanding for the employees; for example, new routines, need for education, and last but not least, in some cases a change of thought. Many of the store owners and managers interviewed felt that it was necessary for someone who was really committed and eager, a genuinely dedicated person, to drive the ecological aspects and issues within the store. They also emphasized the importance of total internal commitment among the employees. One store owner felt that he had to engage himself and his family in the ecological issues first, before he could engage his employees. He therefore started to sort his own household waste and take the refuse to the recycling station before he implemented the eco-strategy in the store.

Within all stores one person was responsible for the internal gathering and spreading of information about ecological issues and products. In some stores they had put together an 'eco-team', including representatives from different departments, which was responsible for planning and implementing the environmental decisions.

Many of those interviewed had tried to influence the operations of earlier channel members in an environmentally protective direction. They had tried to persuade the producers to develop more environmentally friendly products, especially within the chemical–technical assortment but also KRAV-labelled food products. Some store owners and managers thought that many manufacturers were rather passive or reactive in promoting their ecological product alternatives. For example, one store owner had asked a representative of a major national food manufacturer for a discount on one of its KRAV-labelled products in order to stimulate sales, but had not received any positive response from the food manufacturer. Another way of trying to influence or encourage earlier channel members is to give awards for environmentally concerned behaviour. For example, Hemköp appoints the 'ecological supplier of the year' among its suppliers. This supplier must have raised the environmental standards of their products or have developed new products that have less negative impact on the environment.

Many store managers claimed that they are against patronizing consumers and have therefore chosen to offer both KRAV-certified and conventional products within the same product groups to let the consumers choose for themselves. Some interviewees mentioned that they had got rid of the environmentally dangerous alternatives, among detergents for example, as soon as there was an eco-labelled alternative. Some store owners and managers thought they had a heavy responsibility, especially since they had the possibility of influencing the purchasing behaviour of thousands of consumers.

Some interviewees had entered into contracts with local ecological farmers and thought that stores ought to buy from local farmers when possible. One store owner had contacted an ecological market trader and let him place his market stall outside the store every Saturday.

In all stores there were a lot of recycling activities. Paper and plastic were recycled; bags were made from the plastic and returned to the stores. Some stores had found farmers who were willing to accept leftovers of fruit and vegetables as feed stuffs for their sheep and pigs, and one store had found a kennel to receive all meat offal. In some of the stores they conserved oil and fat from the grill and returned it for recycling.

7.5.4 The role of stakeholders' response

All interviewees emphasized the importance of customers' responses to the ecological strategy and activities of the store. Environmentally engaged customers were thought to encourage the store owners and managers as well as the employees to continue their activities within the ecological field. If customers also responded to the actions of the store, by buying KRAV-certified products and/or discussing ecological issues with the employees, this was felt to be a driving force for the store owners and managers to continue their ecological concern.

The actions of competitors were also regarded as important by those interviewed, in that competitors worked as both driving forces and triggers. It was perceived important to keep the same pace as competitors as regards ecological activities, and preferably to be ahead of them. It was, however, not regarded as an advantage to be too far ahead of competitors. Among the store owners and managers interviewed, there was a fear of being too pro-active, if competitors were more or less inactive. The interviewees thought they would lose customers if, for example, they stopped selling environmentally hazardous products that were still offered by their competitors, or that the customers would perceive their store as expensive, because it had a larger assortment of (more expensive) ecological products than the competitors.

Local environmental groups may work as both discussion partners and prompters. For example, once every year the Swedish Society for Nature Conservation arranges, both at a local and a national level, the 'eco-store of the year' contest. Winning this award was perceived as a great honour by the store owners and managers interviewed.

7.5.5 Conclusions

One main conclusion from this analysis is that store owners and managers regarded sales of ecological products so important that they have redirected shelf space from non-ecological to ecological products. Strategies for shelf management have been developed, and much of the efforts of the store managers are directed to the evaluation of different practical approaches to handling the ecological products. The reasons for selling ecological products have been closely connected to a wish to portray a better image, to be perceived as ecologically responsible.

This positive image-building effect, especially among young people, has been clearly noticed by store owners. It seems that the perception of the importance of ecological aspects is quite different among the young, and the cynicism of middle aged marketing people is not at all appreciated by the younger generation.

Thus there are two mutually reinforcing dynamic forces interacting in the retail–consumer interface. One is the eco-strategy and marketing of the retailers: the other is the rapidly growing environmental consciousness among the younger generation. It seems that retailers who can integrate these two forces in the store management, product range and marketing can improve profits.

7.6   The role of government

According to Porter (1990, pp. 126 and 128)

> Government's real role in national competitive advantage is in influencing the
> four determinants. ... Government has an important influence on national com-

petitive advantage, though its role is inevitably partial. Government policy will fail if it remains the only source of national competitive advantage.

In this section we discuss two major ways in which the Swedish government influences the four determinants; through legislation and subsidies.

7.6.1 Government legislation

The 10% goal. In order to reduce the negative environmental effects from agriculture, the Swedish government in 1995 set as a target that 10% (278 000 hectares) of all Swedish agricultural land should be ecologically cultivated by the year 2000. In 'the action plan for the advancement of ecological production', the Swedish Board of Agriculture (1996) defined the 10% target as one that ought to be reached within each branch of production, that is 10% of all grain, vegetables, fruit, dairy products, meat products, etc. should be ecologically produced by the year 2000.

Legislation on food safety and animal welfare. The Swedish law on animal welfare is among the toughest in the world and has its roots not only in a desire to minimize the risk of infections but also in animal ethics. Besides regulations focusing on such issues as the use of carcasses in feeding stuffs and the use of antibiotics, Swedish animal welfare legislation emphasizes the well-being and the need for natural behaviour of animals.

According to Stig Widell (1996) at the Swedish Board of Agriculture, the use of carcasses in animal feed stuff has been prohibited in Sweden since 1986. On joining the EU, Sweden was permitted to keep this as a temporary regulation for 3 years. After the bovine spongiform encephalopathy (BSE) crisis in Britain in 1996, France prohibited all use of carcasses, and today all carcasses have to be cremated. Today there are discussions within the EU on whether or not the use of carcasses as animal feed stuff should be prohibited in the future. Outside the EU, in May 1996 Switzerland prohibited the use of carcasses in animal feed stuffs for livestock production.

The use of antibiotics in animal feed stuffs was also prohibited in 1986 (Stig Widell, 1996). This law has also been permitted to continue in operation as a temporary measure for 4 years after Sweden's accession to the EU. During this period Sweden has to show scientific proof justifying a continuation of the law. In 1986 there was also a regulation regarding animal husbandry which included provision, for example, that all dairy cows and cattle (with exceptions of bulls) older than 6 months, and if possible pigs also, shall spend the summer period outdoors, and that laying hens should not to be kept in cages.

Pesticides. According to Anders Emmerman (1996) of the Swedish Board of Agriculture, the use of pesticides in Sweden is small compared to many other European countries. Furthermore, all pesticides have to be approved by the Swedish National Chemicals Inspectorate. In Sweden there is also a programme for reducing the health and ecological risks as well as the use of pesticides. This programme has among other things resulted in a decrease by 65% in the use of active substances, compared to the average use per year in 1981–1985. Emmerman also remarked that the Swedish farmers are well educated in using pesticides and well aware of their negative effects on the environment.

Within the EU there are common rules for the testing and approval of pesticides, and this rule system has begun to be put into practice in Sweden. We do not yet know the results of the application of this new system. On the one hand it might mean that Sweden has to lower its standards; on the other hand, it might mean higher requirements in other countries, which could have a significant effect on the volume of pesticides available. The EU is also developing a pesticides policy in which ecological considerations seem to receive greater consideration (Emmerman, 1996).

7.6.2 Government subsidies

In order to encourage conversion from conventional to ecological agriculture the Swedish government in 1989 granted all farmers with at least 2 hectares of ecological cultivation a subsidy of 700–2900 SEK per hectare per year for a maximum of 5 years. The land was to be cultivated in accordance with the rules of the Swedish Board of Agriculture. These rules were much the same as those of the Swedish certification organization of ecological production, KRAV. Subsidies were given for grain, oil seeds, peas, beans, potatoes and sugar beet. In the first year the subsidy was also given for pasture land, green forage and green manuring. In 1994 an area-based income support was given to arable land that was KRAV or Swedish Demeter approved, including conversion acreage. The subsidy was 156 SEK per hectare. In 1995 the European Commission improved on the Swedish programme for environmental support for the period 1995–1999, with a subsidy of 900 or 1600 SEK per hectare per year depending on location. An additional 600 SEK per hectare is given for pasture land if the farm includes ecological husbandry (Swedish Board of Agriculture, 1996).

7.6.3 Conclusions

Tough legislation has created selective disadvantages for Swedish food producers. This has also created a trigger for development of production methods other than conventional farming. As a result, and supported by incentive programmes, ecological production methods have rapidly been developed.

The media debate about the tough legislation and Swedish environmental policy has attracted attention from consumers, and the demand for ecological food has been stimulated. At the same time, Swedish local communities have devoted much energy to the fulfilment of the Agenda 21 goals from the Rio conference. Among other things, schoolchildren have had substantial training in ecological and environmental thinking.

7.7 Supporting and related industries

In this section we briefly describe the interesting development of a supporting industry for the ecological food sector. This supporting industry is a knowledge-based industry that provides services and products of great importance for the strengthening of the ecological product concept. This knowledge-based industry develops methods and techniques that will raise the environmental awareness and concern in general among different actors along the food chain, from farmers to consumers. Even though some of these methods and techniques do not focus on ecological production *per se*, but rather take a broader view of environmental issues, ecological food production is certainly supported by this 'environmentally based industry'.

7.7.1 Eco-auditing and certification

Environmental audits. One such product is environmental audits. Some consultancies have developed an environmental audit concept that consists of an audit guide with checklists, describing what factors and what procedures to measure and register to get a good view of the eco-management of a farm. The package also includes instructions on when to measure, and the consultancy provides a control system. Results are sent in and an auditor visits the farm regularly. With the audit pack also goes training programmes and self-evaluation tests for the farmers. One of the key players in this area is the consultancy firm of the national farmers' co-operative organization (LFR Konsult). The national organization has communicated to its members the importance of taking responsibility for the environmental qualities of their own farm, since they have launched the national campaign for more environmentally conscious farming. Therefore, and because farmers have appreciated it, the Environmental Audit has been widely accepted.

KRAV, the certification organization in Sweden. KRAV is an example of an organization that has been institutionalized as a certifying organization in very short time. KRAV was established in 1985 on the initiative of ARF, Samodlarna Svea, the Swedish Demeter and Förbundet Organisk-Biologisk Odling, and is now the leading certification organization in Sweden. The background to its establishment was the need for an independent control of ecological production as ecological products began to be sold in traditional grocery stores. Previously when consumers bought directly from the ecological farmers there was no need for this since much of the control was based on mutual understanding and consumers' confidence in the farmer. In selling ecological products through grocery stores the direct contact between consumers and farmers was lost, hence the need for an independent control of production. KRAV is accredited by the International Federation of Organic Agriculture Movements (IFOAM) Accreditation Programme and is authorized by the Swedish state to control ecological production according to the EC regulation 2092/91.

Today, KRAV has 20 full-time inspectors who control the whole production and distribution chain of ecological production and products, from farmers to retailers. More than 40 employees are involved with standards, certification and information on ecological production. GroLink, a subsidiary organization of KRAV formed in 1995, develops certification programmes outside Sweden (Gardfjell, 1996). At the end of 1995 there were some 350 food-processing companies certified by KRAV, with a range of 1500 products including baby food, juices and jams, dairy products, crispbreads, grain flakes and beer (KRAV guiden, 1996). Retailers are also certified by KRAV, and restaurant authorization also has recently been developed.

7.7.2 Eco-research

Measurement instruments and test sample handling procedures. In order to secure quality in agricultural production, test procedures are being developed to reveal product characteristics that could endanger health and quality. In this development there is an interesting trade-off between measurement results and measurement cost. It is possible to measure almost everything in a production chain, but doing so would make products very expensive and probably impossible to sell. It is therefore important to know what tests are necessary in order to guarantee high quality products, and where in the value added chain these tests should be carried out. This must be based on a thorough knowledge of how different substances are transmitted from one step of the chain to another, and what effects these substances have on the consumer. This quality and health control was formerly left to the authorities (Livsmedelsverket), but today key actors in the food industry invest in the development of this knowledge. One of the important companies is AgroLab, but the major production companies as well as some retailers have their own programmes for quality development of this kind. It should be noted that these health controls are being carried out in primary production, in food processing and in transportation.

Development of alternatives to pesticides. A very interesting part of the eco supporting industry is the development of alternatives to pesticides. Crop species have been developed that can resist various kinds of vermin and climatic conditions. This will create plants and crops that are less sensitive and can grow without too much support. Many of these new products are still in the R&D phase in companies such as Svalöf-Weibulls and at the Swedish Agricultural University, but these plants and crops will be on the market in the future. In another development, researchers are seeking to replace traditional plants and seeds with others that can produce better production results. This is especially valid in the production of industrial products. A third development concerns biological insect control. In this field, instead of using pesticides to kill insects, you find other insects that will eat the ones that harm the crop, or you stimulate the interest of the insects with the help of biological means to do other things than to eat the crop.

7.7.3 Eco-knowledge development

Consumer knowledge development. There are three major actors who actively teach the population about food and health. This information is not only about ecological food, but, since quality and heath are the key issues that are being taught, it will to some extent strengthen the ecological products. If the eco product sellers link to the messages that children have learned in school, they can reach customers with their arguments. The key actors in this education represent the state (Livsmedelsverket), the food processing companies (Livsmedelsbranschen) and the farmers' co-operative (LRF).

7.7.4 Conclusions

In this section we can see the emergence of a knowledge based industry. This industry develops specific knowledge and specific skills and unique products that all have the purpose of making the quality of ecological production higher. At the same time the effort is to make the production, distribution and sales of ecological products more efficient and cost effective. This industry will play an important role in the dynamics of the ecological industry and will contribute to increased competitiveness of ecological products.

It is quite interesting that the knowledge developed in this industry is valuable in itself. The knowledge and products developed here can be applied not only in ecological production and sales, but also in the production and sales of other high quality food products. It is also interesting that, from an internationalization point of view, this knowledge-based industry can – if it has unique products – sell its services in foreign markets even if not a single Swedish ecological product goes into that market. Here we can also see openings for European co-operation between ecological producers, in the form of benchmarking and exchange of best practices.

7.8   The competitive dynamics of the Swedish ecological food cluster

The last part of the Porter (1990) model that we want to mention briefly is **chance.** Looking back at recent Swedish exports of food with high health qualities, we note that the ability to take advantage of chance events is important when it comes to commercial exploitation of high quality food concepts.

One such example is the starting of exports of ham and salmonella-free chicken to Denmark. In the ham case, health related production problems and strikes in slaughterhouses in Denmark were the chance events that created opportunities for Swedish exports. In the chicken case, a strategic change in a Danish consumers' co-operative, resulting in the promotion of ecological food using lower retailer margins, was of importance. These events created business opportunities for Swedish products in Denmark, even though Denmark is a sophisticated country when it comes to ecological production and food quality in general. Just as when we discussed demand above, chance events focusing on food safety and food quality tend to strengthen the market position for ecological alternatives.

A chance event that also has accelerated the process is the British BSE catastrophe, which really has communicated to the consumers that you should be aware of what you eat. These insights do not lead the consumer to be interested in the details of the scientific debate on what causes what, when and how. From a consumer point of view it is enough to know that ecological food seems less dangerous.

Adding this aspect to the others analysed above, we get a full picture of the interaction between the different parts of the ecological food cluster. We can identify the triggers and the driving forces in the dynamic process that creates competitiveness.

The major retailers have developed ecological strategies that result in large volumes of ecological products in the stores and a strengthening of the ecological messages in their marketing. The efficient distribution and marketing by the retail chains helps larger volumes and well designed sales messages to flow through the value added chain. The combination of retailers' private brands and KRAV certificates creates a strong concept. We also notice an important rivalry aspect. Since the three major players in the retail market all have ecological strategies, the ambition to implement this strategy better than the main competitors creates a driving force in the upgrading process.

When the retailers launch their ecological strategies, they are met by a market that is in agreement with the main themes. The health and environmental awareness in the population has been increased through the media debate on the legislation issues, through schools and through nationwide media and education projects carried out by authorities, the food industry, the farmers' co-operative and the retailers themselves.

The government have through legislation created selective factor disadvantages for traditional bulk food production. The positive force coming out of this is that in doing this, they have driven the ecological producers into development of efficient production factors and methods, a newly designed distribution network and a certificate that has been institutionalized so that the consumers trust the label. Production subsidies promote ecological farming.

A supporting industry is in the course of development. In this industry, methods, techniques, products and services are developed and offered to the market. These products and services are all directed to the strengthening of the ecological product concept in various parts of the food chain. Since the production volumes are now increasing, it is increasingly profitable for small niche suppliers in this service industry to develop specialized knowledge and invest in specialized equipment. The number of customers is also growing. These products and services have a value in themselves, and can be used outside the ecological food cluster.

The Swedish case illustrates how the different parts of the Porter diamond support each other in the development of increasing dynamism and competitiveness. We have illustrated the importance of 'market fit' and 'market timing', referring to a fit between supply and demand that is encouraging farmers to convert and manufacturers and retailers to develop and market ecological products, as well as timing between the rate of certification of farmers, important intermediaries, manufacturers, stores, etc. We have also emphasized the importance of agreement of a certification system as well as an engagement in environmental issues throughout the food chain as a whole. Furthermore, we have illustrated that changes in one or a few factors may give rise to both virtuous and vicious circles; for example, proactive retailers may stimulate demand for ecological products as well as encourage manufacturers and farmers to develop and produce ecological products.

In the next section we will relate this analysis to conditions in other European countries.

7.9. Anticipatory buyers' needs

Ecological production is on the increase in many European countries (Figure 7.11). An official EU publication (Green Europe, 1994) mentioned a market share for ecological products of about 0.5% of the total market for the EU as a whole, and that a forecast of around 2.5% seemed reasonable for the year 2000. This forecast did not, however, include Austria, Finland and Sweden. Since the share of ecological agricultural land (Table 7.3), at least in Austria, is very high (20%), the share for ecological products for 2000 may be assumed to be even greater than 2.5%.

7.9.1 Some competing ecological food clusters

Austria. The Austrian alpine landscape makes large scale farming more or less impossible. Many farms are small, and farming is combined with pensions or catering for tourists to provide adequate incomes. In the alpine areas the cold winters also contribute to a low use of pesticides, since vermin die during the winter period. Traditional farming in Austria is thus close to

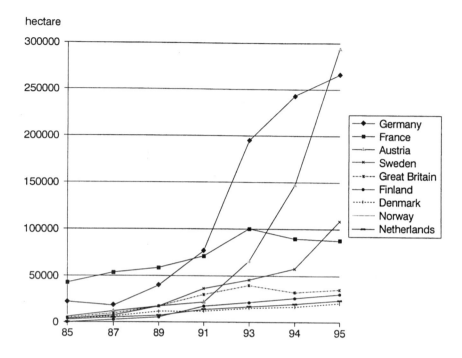

Figure 7.11 Ecological production in some European countries (hectares) (KRAV kontrollen, 1995).

being ecological anyway, which makes it easy to convert. Mostly farmers have only to adopt ecological animal housing standards. Since Austrian farming cannot compete in low cost production, 'Ecoland Austria' became the strategy of agricultural policy in preparing for EU membership in 1995. Today all major retailers sell ecological products and they have also developed their own ecological private brands, for example, the market leader Billa intro-

Table 7.3 Share of ecological cultivated and convertion acreage in some European countries 1985–1995, percentage

	1985	1987	1989	1991	1993	1994	1995
Austria	0.7	1	1.3	1.7	4.7	9.5	19.4
Denmark	0.2	0.2	0.4	0.7	0.7	0.8	1.1
Finland	0	0	0	0.6	0.8	1.0	1.1
France	0.2	0.3	0.3	0.4	0.5	0.4	0.4
Germany	0.2	0.2	0.3	0.6	1.6	2.0	2.2
Netherlands	0.2	0.3	0.7	0.9	1.0	1.0	1.3
Norway	0	0	0	0.6	0.8	0.7	0.8
Sweden	0.2	0.3	0.4	1.4	1.6	2.1	3.6
UK	0.1	0.1	0.3	0.5	0.6	0.5	0.5

Source: KRAV kontrollen, 1995.

duced Ja, Natürlich in 1994 and in the following year the number two retailer Spar followed with Natur Pur. Ecological products are labelled with the single Bio control mark (Thimm, 1995).

Denmark. In Denmark the share of ecological production, especially of milk production, has increased during the last few years (Borgen, 1995). Different ecological producer co-operatives have been established during the last 15 years. In their 'Action plan for the promotion of organic agriculture' the Danish government anticipated that by the year 2000, 7% of its agricultural land would grow ecological products. In 1987 Denmark adopted legislation on ecological agricultural production, and in 1988 the Danish government began granting subsidies to ecological production. Ecological products are sold in almost all supermarket chains. As in Sweden, a broad assortment of ecological products is an important part of the retailer profile. The largest of the retailer chains, FDB (co-op) was first to sell ecological products in its stores in the early 1980s. It now sells about 400 different ecological products, and ecological milk accounts for 15–20% of the chain's total milk sales. Today 70% of all ecological products are sold through major traditional food retailer stores. Ecological products are marked with the red control mark Ø, a symbol on which Danish consumers rely (Thimm, 1995).

Germany. Generous government subsidies together with a supportive market resulted in conversion of many conventional farms in Germany (KRAV kontrollen, 1994), However, the share of total arable land is still very small, only 2.2% in 1995 (Table 7.3). Only a small proportion of ecological products is sold through the conventional food retailers: most are sold in speciality stores (Buley, 1995) or at the weekly markets (Hagman, 1995). Since only about 3% of the total production of vegetables in Germany is bought at the weekly markets compared to 80% in supermarkets (Hagman, 1995), it can therefore be assumed necessary to increase the sales through the supermarkets in order to achieve a large volume distribution. In Germany there is a huge number of independent control bodies (about 50) controlling the actors in the food chain. The certification of farmers is made by eight farmers' associations that are accredited according to EU regulations. (Buley, 1995)

The Netherlands. Some 60–70% of ecological products are sold through specialized natural food stores and about 20% through supermarkets in the Netherlands (Jeuken, 1995). Specialized food stores are a stable sales channel, neither expanding nor diminishing in sales, expansion is mainly caused by the supermarket chains. The interest from the supermarkets is, however, constantly changing and it is doubtful if their share will increase as much as in other European countries (Hagman, 1995). One reason for this might be that retailers have so far favoured integrated agriculture production (Jeuken, 1995). In the Netherlands there is one independent certifying body,

Skal, which in addition to the EU standards has developed standards for animal husbandry and production (Jeuken, 1995).

Belgium. There is a slow growth of ecological production in Belgium (Boulanger, 1995). One reason for this might be that the ecological production is scattered and there are difficulties in grouping the supply (Hagman, 1995). The produce is sold through three different channels; at the public market place or farm sales, through natural and health stores, and through supermarkets. There are two control organizations accredited by the state, BLIK and ECOCERT-BELGIUM, which also certify animal-derived products (Boulanger, 1995).

UK. In the UK the growth of ecological production is slow or even stagnant (Table 7.3). About two-thirds of ecological products are sold through supermarkets such as Sainsbury's (Hagman, 1995). In 1994 the government introduced an Aid Scheme which granted assistance to farmers during conversion to ecological production (Marriage, 1995).

France. There is no growth of ecological production in France either (Table 7.3). About a quarter of ecological products are sold through health food stores, a quarter through supermarkets, and a third through farm sales or direct sales. There are two state accredited control bodies, ECOCERT and QUALITÉ-FRANCE (Sylvander, 1995).

Spain and Italy. In both Spain and Italy the export market for ecological products is more important, and growing, than the domestic market. There is a very limited interest from the conventional distribution system in handling ecological products, and supermarkets are largely unrepresented. In Spain the domestic market is more into diet and health products than ecological products (Picazos, 1995), and in Italy there are problems of positioning ecological products versus integrated production (IP) products (Haest, 1995).

Switzerland. In Switzerland the major part (about 55–60%) of ecological products are sold through supermarkets, COOP and MIGROS, 25–30% is sold at specialized stores and the rest at farm sales. However, MIGROS did not enter the ecological market until 1994: before then it was very active in IP. Since 1993 government subsidies have increased the area of ecological pasture land, particularly in mountainous regions (Knauer, 1995).

7.9.2 Conclusions

There are many similarities and differences between European countries regarding ecological production and products. Some of the main factors which have been described above are (cf. Hagman, 1995):

- the process of conversion
- the structure of the supply
- the character of the distribution system
- the number of certifying organizations and control marks
- the support by government.

How difficult or easy it is to convert is dependent, for example, on the characteristics of traditional agriculture production and the financial position of the farmers. Basic factor conditions, such as climate and topography, may have had a major impact on the traditional production methods. In the alpine areas (most of Austria and Switzerland, parts of France and Italy) and the northern Scandinavian countries (Sweden, Norway and Finland) there is already in traditional agriculture a relatively low need for pesticides, since most vermin die during the cold winters. In the alpine areas large-scale agriculture is more or less impossible becasue of the topography, which has forced people to combine farming with running guesthouses, etc. Becasue they have incomes from other businesses, it may be assumed that these farmers are not as risk averse as their counterparts in the rest of Europe. In the Netherlands, on the contrary, agricultural production is utterly intensive: here we might assume that it is easier to convert to IP instead of ecological production methods.

The more scattered the supply, the harder it is to sell ecological products through large-scale conventional distribution systems, such as supermarkets and traditional food stores. In many countries, producer co-operatives have been developed with the aim of gathering the supply from many different farmers into large enough quantities to supply the conventional food chains. It may be assumed that the more concentrated the supply of ecological production the easier it is to mass market the products. Furthermore, it may be assumed that the higher the share of ecological production sold through large-scale distribution systems the lower the price to the consumer, and the more motivated the consumers are to buy ecological products.

If ecological products are only sold through specialized food, natural or health food stores, it seems reasonable to believe that these products never reach the non-devoted consumers (consumers not specifically concerned with health and environment issues). In order to reach the majority of consumers, ecological products must be available in traditional food stores. In the case of Austria, Denmark, Sweden and Switzerland, the market has developed mainly due to the large quantities sold through traditional food stores and supermarkets. In these countries it is also common to find retailers that have developed their own ecological private labels.

The more numerous the different inspection bodies and control marks on the market, the harder it is for consumers to learn to distinguish the ecological products or as Tate (1994, p. 16) writes:

> The more certifying bodies there are operating independently of each other in one country, the greater the uncertainty in consumers' minds and the greater the requirement for the introduction of a common national standard.

In Austria, Belgium, Denmark, France, the Netherlands and Sweden there are only one or two certifying bodies and eco-labels on the market, whereas in Germany, the UK, Italy and Spain there are many different labels.

Governmental support for ecological production, either by subsidies or by national policies, differs from country to country. In Austria, Denmark and Sweden the government has developed action plans for ecological production and explicitly claimed that conversion to ecological agriculture is desirable. In Germany the government has also been generous with subsidies regarding conversion and ecological production.

Furthermore, free and public market information regarding prices and quantities is a major factor distinguishing between countries (Lampkin and Padel, 1994; Hagman, 1995):

> However, the communication of existing information is perhaps the most important factor which would reassure farmers that organic farming is a viable option, reducing the role of lack of confidence as a barrier to conversion, and which would ensure that the potential for negative impacts of conversion is minimized. Germany, Denmark, Sweden, Finland, Norway, Switzerland, Austria and the Netherlands have full-time state-funded advisers specializing in organic farming. (Lampkin and Padel, 1994, p. 451)

7.10   Concluding remarks

At this stage in the analysis, we ask ourselves if the increasing competitiveness of the Swedish ecological food cluster in the national market could create a base for international competitiveness. In one sense, and according to our definition presented earlier, this international competitiveness is already a fact, in that the strength of the Swedish ecological concept has created growth and market share in the national market – market shares that have been taken from traditional food concepts, but also from ecological food from other countries.

In our view, the following conditions have to be met if Swedish ecological food cluster are to achieve success in other markets:

- If the conclusion that the ecological awareness is growing in younger generations, is valid also in other markets, the market conditions necessary will exist. This would create a large market segment, at least on a European base.
- If efforts to secure the supply of high quality ecological products throughout the year are successful, this will strengthen the cluster. In order to achieve this, it is important to realize that Sweden is too small a base for a European market. Co-operation with strong producers and distributors in other countries could be a way to success. This kind of co-operation could be stimulated by the fact that the strong ecological

producing countries in Europe have different foci of production. For this reason, products from different countries can be complementary and not primarily competing.

- If efforts to institutionalize ecological production and product certification on a European level are successful, this will help the process. The estimated success rate will increase if ecological producers in Europe recognize that the market is growing and that the main competition is not ecological products from other countries, but traditional food products from all European countries.

- If the supporting industry is successful in developing instruments, measurements, logistics, procedures and management capabilities for the continuous upgrading of the cluster abilities in every section of the value added chain, this will be of major importance for the success of the ecological product concept on a European level.

- If the production and distribution systems have the ability to increase cost effectiveness when the volume increases, the prices of ecological products will decrease in comparison with other food products. This will result in higher volumes and market shares.

Looking at the capabilities that are developed in the process described above, we discover that many of the strengths and skills needed to be a competitive ecological retailer are the same as those needed for all retailers with a strategy to market high quality niche products and concepts in national and foreign markets. We would therefore like to propose that the knowledge and experience created in the ecological food cluster, could be transferred to other niche product strategies.

This transfer of knowledge can be created in four parts of the Porter diamond:

- **Supporting industries**: The knowledge-based supporting service industry that produces ecological measurement techniques, ecological audits and ecological production methods will develop a competence base which can be used for production of other high quality food and value added products. In this way, the specialized and advanced knowledge in this industry can be a driving force in the creation of other high value added products.

- **Factor conditions**: The development of new species of plants that will be more resistant to vermin and weather conditions can improve productivity in the production of other food products. Farms with ecological experience will develop a more refined approach to quality control that can be used even if they change over to other production concepts.

- **Strategy and structure**: The development of a specialized organization for the collection of products from many producers and channelling it into the major retail channels will create a learning effect in the industry. This may create an increased ability to organize the industry for the efficient distribution of other niche products.

- **Government**: The basic thinking in the competition legislation concerns basic products and their price competition. From the ecological area we will learn that quality and price are both important variables, and that building of product quality will in some cases be a result of co-operation between the different parts of the value added chain. Furthermore the ecological experience will be useful in the revised legislation for other products that can be proven healthy in some respects – functional food. The ability to create legislation for products that will improve health, but should not be regarded as pharmaceuticals, will be upgraded if experience from the ecological field is capitalized upon.

In this way the ecological food cluster can contribute to the upgrading of other parts of the food chain.

This case has illustrated the importance of ecological commitment throughout the whole food chain in order to generate dynamism, within the diamond that will raise standards within the industry. Yet, we would like to suggest that the ecological responsiveness of Swedish food retailers may be regarded as a driving-force for that dynamism.

Not only have the retailers encouraged or forced the supply side to co-ordinate and gather the produce, they have also taken an active part in the development and marketing of ecological products. The proactive eco-strategies of the major retail chains as well as store owners and managers have also created valuable knowledge and experiences about retail niche marketing and image-building among the retail actors.

It is now time to ask not only what the food industry can do for the eco-logical food cluster, but also what the ecological food cluster can do for the food industry.

7.10.1   The usefulness of the Porter framework

In this study we have used the Porter (1990) framework in order to forecast future developments. The framework has been useful in a sense that it not only make us analyse different factors which might explain future competi-tiveness, but also forces us to analyse how the different factors are inter-related and reinforce each other. In this respect the diamond has been most useful. Because of its broad approach, the analysis of each single factor might be less deep than if only one or two factors were focused. However, the usefulness of the framework, as we see it, is not in the detailed analysis of each and every factor, but rather that it forces researchers to consider the dynamism among the factors within the diamond. Especially for forecasting purposes, we believe it is more important to understand what is driving the development, and how one or a few factors may give rise to both virtuous and vicious circles for future competitiveness, than to analyse all aspects of one or two factors in detail. Whether or not the framework has been fruitful in the sense that it has made it possible to give a 'true' picture of the future, only time will tell. We have tried to use the framework in order to indicate the conditions that have to be met if the Swedish ecological food cluster is to become a competitive industry with the sustained ability to profitably gain and maintain market share in domestic and/or foreign markets (Agriculture Canada, 1991).

References

Articles, books and reports

Agriculture Canada (1991) Task Force on Competitiveness in the Agri-Food Industry: *Growing Together: Report to Ministers of Agriculture*. Agriculture Canada, Ottawa.

Borgen, M. (1995) *Økoguide 1995*. Økologisk Landcenter, Denmark.

Boulanger, M. (1995) Belgium, in *International Organic Market Study* (ed. C. Haest), BioFair, Costa Rica.

Buley, M. (1995) Germany, in *International Organic Market Study* (ed. C. Haest), BioFair, Costa Rica.

EC (1991) Council Regulation No. 2092/91 of 24 June 1991 on organic .production and indications referring thereto on agricultural products and foodstuffs. *Official Journal of the European Communities*, 91(L198), 1–15.

Gardfjell, M. (1996) 10% organic by 2000: the official Swedish goal. *Ecology and Farming*, 12, (May–August), 19.

Green Europe (1994) *Organic Farming* No 2. European Commission, Brussels.

Haest, C. (1995) Organic Italy, in *International Organic Market Study* (ed. C. Haest). BioFair, Costa Rica.

Hagman, T. (1995) Decision and location factors in organic production of vegetables. MSc Thesis, SLU, Alnarp.

Holm, H. and Drake, L. (1989) *Konsumenternas attityder till alternativt producerat kött*. (Consumers' attitudes towards alternatively produced meat). Institutionen för ekonomi, Uppsala.

Jeuken, M. (1995) Holland, in *International Organic Market Study* (ed. C. Haest). BioFair, Costa Rica.

Knauer, K. (1995) Switzerland, in *International Organic Market Study* (ed. C. Haest) BioFair, Costa Rica.

Krav-guiden (1996) KRAV, Uppsala.

Krav-kontrollen (1994) KRAV, Uppsala.

Krav-kontrollen (1995) KRAV, Uppsala.

Krav-regler (1996) KRAV, Uppsala.

Lampkin, N.H. (1994) Organic farming: Sustainable agriculture in practice, in *The Economics of Organic Farming. An International Perspective* (eds. N.H. Lampkin and S. Padel). CAB International, Wallingford, Oxon.

Lampkin, N.H. and Padel, S. (1994) Organic farming and agricultural policy in western Europe: an overview, in *The Economics of Organic Farming. An International Perspective* (eds N.H. Lampkin and S. Padel). CAB International, Wallingford, Oxon.

LUI/Samodlarna Sverige (1994) Konsumentundersökning om KRAV-märkta livsmedel – Hur ser den ekologiska konsumenten ut? (Consumer survey of KRAV-labelled food products – Who is the ecological consumer?). Stencil.

Marriage, C. (1995) Great Britain, in *International Organic Market Study* (ed. C. Haest). BioFair, Costa Rica.

Mathisson, K. and Schollin, A. (1994) *Konsumentaspekter på ekologiskt odlade grönsaker – en jämförande studie*. (Consumer aspects of ecologically grown vegetables – å comparative study). SLU, Uppsala.

Mathisson, K. and Schollin, A. (1995) *Ökat intresse för ekologiskt odlade grönsaker*. (Increased interest in ecologically produced vegetables) Trädgård no. 4, SLU Fakta, Uppsala.

Picazos, J. Spain, in *International Organic Market Study* (ed. C. Haest) BioFair, Costa Rica.
Porter, M. (1990) *Competitive Advantage of Nations*. Macmillan, London.
Sahlström, K. (1983) Alternativa odlingsformer, in *Nya matvanor* (New eating habits) (ed. B. Nielsen, Grupptryck, Miljöhäfte 15–16, Bokskogen 34–37, Falun.
SAO and Bokskogen (1983) *Biologisk odling. Odlingssystem med helhetssyn* (Biological farming. Farming system with a holistic perspective). Grupptryck, Lindsberg, Falun.
Svensk Dagligvaruhandel 1995–1996. Fri köpenskaps Förlag, Solna.
Svensk Dagligvaruhandel, 1993–1994. Fri köpenskaps Förlag, Solna.
Swedish Board of Agriculture (1996) *Ekologisk produktion – Aktionsplan 2000* (Ecological production – Action plan 2000) Report No. 1996:3, Jordbruksverket, Jönköping.
Sylvander, B. (1995) France, in *International Organic Market Study* (ed. C. Haest). BioFair, Costa Rica.
Tate, W.B. (1994) The development of the organic industry and market: An international perspective, in *The Economics of Organic Farming. An International Perspective* (eds. N.H. Lampkin and S. Padel). CAB International, Wallingford, Oxon.
Terrvik, E. (1995) Implementering av miljöstrategi genom eget miljövarumärke – Fallet Änglamark. (Implementation of environmental strategy through own eco-brand – The case of Anglamark), in *Miljöstrategier – Ett företagsekonomiskt perspektiv* (Environmental strategies – a business economical perspective) (eds I.P. Dobers and R. Wolff), Nerenius and Santérus Förlag, Göteborg, pp. 109–28.
Thimm, C. (1995a) Austria, in *International Organic Market Study* (ed. C. Haest). BioFair, Costa Rica.
Thimm, C. (1995b) Denmark, in *International Organic Market Study* (ed. C. Haest) BioFair, Costa Rica.
Wennerholm, T., Lagnevik, M. and Göransson, G. (1996) *Scenarioteknik vid strategisk planering – tillämpningar inom livsmedelskedjan.* (Scenario technique during strategic planning – application to the food chain), LOK-rapport No 7, Department of Business Administration, School of Economics and Management, Lund University.

Company internal documents and external brochures

Eco Trade (1995) Market information (newsletter) 5 September 1995.
Ekokött (1996) Sales statistics (internal document)
Gröna Konsum (1996) Sila Kamelerna (Look after the big issues and let go of the smaller ones).
Hemköp (1995) Hemköps Miljöpolicy (Hemköp's environmental policy)(internal document)
ICA Handlamas AB (1997) Environmental Policy.
KF (1994) KF Miljörapport (Environmental report)
KF (1996) Verksamheten 1995.
Telephone interviews with representatives of authorities and producer co-operatives
Edh, G., Eco Trade, 22 October 1996.
Emmerman, A., Swedish Board of Agriculture, 22 August 1996.
Hacklou, B., ICA Frukt and Grönt, 26 August 1996.
Thulin, Y., Mejeriindustriernas Förening, 8 October 1996.
Widell, S., Swedish Board of Agriculture, 22 August 1996.

8 New policies, new opportunities, new threats: the Finnish food industry in the EU

SAARA HYVÖNEN AND JUKKA KOLA

8.1 Introduction

8.1.1 EU membership: new challenges for the Finnish food chain

The major change for the Finnish food chain was membership of the EU in 1995. Although Finland is an open economy, before 1995 60% of the domestic food industry belonged to the closed, protected sector of the economy with hardly any foreign competition. Some leading co-operatives were in practice even capable of preventing new domestic firms from entering the market because they controlled the sources of raw materials. In the Finnish food chain, the wholesale and retail sector is highly concentrated. The two co-ops and the two quasi-integrated distribution channels account for 95% of all food sales at the retail level. Buyer concentration restricts channel alternatives open to food manufacturers, and weakens the dynamics of competition throughout the whole food chain (Tirole, 1988). Agricultural and food policy have mainly been based on the principles of self-sufficiency and protectionism (Kola, 1993). Consequently, a significant part of the food industry has formulated its strategies under the assumption of continuing stability and government subsidies, resulting in an orientation of 'sell what we can make' (production-led) rather than of 'make what we can market at a profit' (market-led). Selling of what had been produced was usually based on price competition in mass markets (Hyvönen and Kola, 1995).

Nowadays the open but more uncertain single European market (SEM) introduces and requires substantial changes in both public policies and business strategies in Finland. In the short run, however, the ability to adjust to the changing environment of the future may be quite limited in the agriculture and food industries owing to sector- and country-specific characteristics such as seasonality and peripherality. In the old situation certain characteristics were also maintained by public policy decisions, and as they often appeared disadvantageous and produced inefficiencies, due to very small-scale structure for example, they were then compensated for by public spending. Public policy did not promote structural changes and improvement in competitiveness based on relative advantages. Now, public policy should play a more active role in creating a business environment which enhances the ability of firms operating in the food chain to compete successfully in the long run.

8.1.2   Objectives of the study

Our study aims to identify the key factors, structures and competitiveness in the Finnish food industry. In analysing competitiveness in the food chain/economy, we employ Porter's diamond model where the potentially beneficial role of government policies is emphasized. From this point of view, our objective is to investigate whether public policy, and of what kind, e.g. agricultural, industrial, competition, trade, finance, or research policy, may alleviate adjustment problems and enhance competitiveness when the political and economic environment changes drastically. Furthermore, we introduce two alternative theoretical approaches for the investigation of competitiveness at the firm level. In terms of Porter's diamond model, the firm-level approach describes strategy, structure and rivalry. Based on data collected from food manufacturing firms, the objective is to develop a taxonomy of competitive advantages. Possible interrelationships between the particular types of advantages and some measures of organizational performance are also explored. A description of the structural characteristics of the food industry is provided, and the study concludes with a discussion of the empirical results as well as policy implications for the competitiveness of the Finnish food chain.

8.2   Structural characteristics of the Finnish food industry

The food industry is the third largest industry in Finland, after the metal and forest industries. In 1994, the major components were slaughtering and meat processing (21% of value added), beverages (16%), bakery products (16%), and dairy products (15%) (Finnish Food and Drink Industries, 1995). The share of the food industry in both the total value added and employment in Finnish manufacturing industry was about 10%. For the whole national economy, they represent about 3% and 2%, respectively. The significance of the food industry in the national economy has remained quite stable (3–4% of GDP since the 1960s), whereas the role of agriculture in the national economy has declined considerably (from 10% to less than 3%). Today, the shares of GDP and total employment of Finnish agriculture, about 3% and 7% respectively, are quite close to the corresponding EU averages. Together, the production input industry, agriculture, the food industry, and related services employ some 300 000 people, i.e. about 14% of the total labour force in Finland.

Finland's membership in the EU had fundamental and immediate effects on agriculture and food production. Farmers had to adjust to the common prices of the Common Agricultural Policy (CAP), which were on average about 40% lower than pre-EU producer prices. The food industry faced increasing import competition. Farmers' loss in income due to the price fall has been compensated to some extent by several national and EU support measures, and the food industry also received some transitional support agreed in the Accession Treaty (Kettunen and Niemi, 1994; Kettunen, 1996).

An important characteristic of the Finnish food industry is the strong role played by farmer-owned co-operative organizations. In 1995, their market shares were 94% in milk, 79% in eggs and 68% in meat. Especially in milk and meat, domestic primary producers and processors strategically depend on each other. These two sectors are also highly concentrated and dominated by a few big co-operatives. Yet the meat processing industry is also fragmented, consisting of about 100 firms which are mainly very small. The close links between the farmers' union, farmer-owned co-operatives and key political powers, i.e. the agrarian-rural Centre Party, have increased farmers' and food firms' political power. Consequently, they have benefited substantially from the strongly interventionist, high-subsidy, and stability-emphasizing agricultural and food policies of pre-EU Finland. As the structure of industry is often used as a proxy indicator of market power, it is expedient to look at the concentration ratios of the Finnish food processing industries. As expected, the value of the concentration ratio has usually fallen when the number of firms has increased, and vice versa (Marttila, 1996). However, in meat processing and in the production of malt beverages and soft drinks, for example, the number of firms has increased but, at the same time, the market position of the leading firms has become stronger, as indicated by the bigger value of the CR3 (Table 8.1). The manufacture of dairy products also became much more concentrated between 1986 and 1993. The Finnish food processing industry is clearly quite oligopolistic, except for the manufacture of fresh bread and pastries, although even the biggest Finnish companies are small in international comparison. Valio Ltd, the largest Finnish dairy company, ranks only about 15th among European firms manufacturing dairy products.

Trade in basic foodstuffs has mainly been in the form of subsidized exports. Trade flows are naturally affected by Finland's geographical location. Food exports to the former Soviet Union were still very important in the 1980s. After a drastic decline in 1990, the proportion exported to eastern European destinations has again increased (Table 8.2). Today, Russia's

Table 8.1 Structure of the Finnish food industry in 1986 and 1993, number of firms and three-firm concentration ratios (CR3)

Industry	Number of firms		CR3	
	1986	1993	1986	1993
Slaughtering	26	26	74.7	71.7
Meat processing	110	127	54.7	60.5
Dairy products	99	67	47.9	65.4
Ice-cream	5	4	99.5	99.4
Beer and soft drinks	15	27	91.4	99.1
Bread and pastries	786	870	38.1	29.9

Source: Marttilas, 1996.

Table 8.2 Agricultural and food trade of Finland by region in the 1990s (%)

Region	Exports						Imports					
	1990	1991	1992	1993	1994	1995	1990	1991	1992	1993	1994	1995
EFTA	15.5	18.0	16.9	15.3	15.7	4.1	16.8	16.7	16.3	14.6	14.3	4.5
Sweden	11.1	13.4	13.2	12.0	12.7	–	8.7	9.1	7.8	7.0	7.2	–
EU	19.5	23.7	25.5	20.5	16.3	38.5	42.1	42.2	42.7	43.0	40.5	53.7
Denmark	3.2	3.6	3.4	3.3	2.1	3.1	6.1	6.5	6.1	5.7	5.8	9.2
Germany	3.4	4.5	6.0	3.7	3.1	5.1	6.2	6.4	6.9	6.6	5.9	8.3
Eastern Europe	37.2	18.8	29.5	41.9	49.6	42.9	2.9	3.1	3.6	3.2	3.4	1.9
Others	27.8	38.0	28.2	22.3	18.4	14.5	38.2	38.0	37.4	39.2	41.8	39.9
USA	10.9	15.1	7.5	10.4	8.1	3.8	5.6	6.0	6.2	6.2	5.6	4.2
Total value, (FIM million)	2358	2241	2640	4134	5162	4042	4995	5208	5885	6975	8540	7703

Source: Finnish Food and Drink Industries.

share is overwhelmingly the largest among individual countries. On the other hand, only a few percent of food imports to Finland come from eastern Europe, but about 40% from the EU in 1990–94, and more than 50% now that Finland and Sweden are in the EU.

Although strong import competition was expected in the open EU markets, it was not yet fully realized in 1995 (e.g. Aaltonen, 1996; Kettunen, 1996). In addition to peripherality, the high degree of saturation and small size of the Finnish market, the immediate disappearance of the gap in food prices between Finland and the EU at the very beginning of 1995 removed major incentives for foreign firms to penetrate the Finnish markets. Finnish food processing firms also made a strategic choice to keep prices at a low level in order to prevent growth in food imports entering Finland. In fact, producer prices of many Finnish agricultural products were the lowest in the entire EU in 1995–1996 (Kettunen, 1996). Finally, one specific reason was the BSE scare in the EU in 1996, which virtually halted beef imports and also slowed down other meat imports to Finland. In 1995, imports represented 7% of meat consumption, 8% of cheese consumption and 14% of yoghurt consumption (Laurila, 1996). Before EU membership, there were hardly any imports of these, or any other basic food products, except some speciality cheeses. In some products, e.g. pig and poultry meat and yoghurt, consumption also increased due to remarkable price decreases. For instance, pigmeat saw a price decline of about 25% and a consumption increase of about 12% in 1995. Overall, food prices fell some 11% in 1995, the first year of EU membership.

Demand conditions in Finland are favourable for domestic food production. Finnish consumers prefer domestic foods to foreign ones, due not only to the perceived quality, safety and freshness characteristics of the products themselves, but also to agricultural production methods that are perceived as sustainable, environmentally friendly and ethically acceptable. Several surveys have indicated this tendency, and the recent food safety and animal disease problems in other countries, e.g. BSE, pig plague and salmonella, have strengthened the domestic preference. In addition, relatively slowly changing national taste habits, although not so different from those of other northern EU countries, and established Finnish brands, provide certain advantages for domestic agricultural production and food processing. Although these advantages may gradually disappear as food imports increase and eating habits in Europe converge, the origin of food will be an important determinant of demand for food in Finland for some years to come.

Major developments in food distribution channels took place in 1996. Two of Finland's four grocery retailers, Kesko and Tuko, worked out a merger deal which envisaged Kesko absorbing Tuko. This merger would have created Finland's biggest grocery retailer, with a domestic market share of 60%, and Europe's 17th biggest retail grocery chain. However, the Finnish com-

petition authorities notified the European Commission of the proposed
merger in summer 1996, and the EU competition authorities refused to
approve the merger in its envisaged form in autumn 1996. Kesko's later offer
to sell 40% of Tuko to new owners did not satisfy the EU, either. The final
solution was still pending in spring 1997. In terms of the relationship
between competition and policy, which will be dealt with in the next section,
these decisions are important. Although EU competition law has applied in
Finland since her accession at the beginning of 1995, its effect here was
unanticipated: as retailers operate on domestic markets, not directly affecting competition in the SEM, the EU was not expected to intervene.

8.3   Theoretical framework

8.3.1   Public policies and the business environment

Competitiveness has become a major concern for governments, as well as
companies, as trade barriers have been reduced between countries (the
SEM; the GATT Uruguay Round agreement; the NAFTA in North
America). Protected agriculture and food sectors face further changes as
policy reforms proceed (eg. the 1987 OECD Ministerial Communiques; the
1992 CAP reform; the 1996 US Farm Bill). Consequently, more attention
has recently been paid to the linkages between certain public policies and
competitiveness of the entire agri-food industry (Brink and Kola, 1995;
Henson et al., 1995; Jones, 1995; Kola et al., 1996).

Good intentions of public policy can – and unfortunately too often do –
result in negative effects, which arise mainly as a result of attempts to isolate
and protect some industry from competition, for example by giving subsidies
and allowing deviations from general legislation (such as competition policy). The infant industry argument is a familiar example. These attempts usually lead to social costs, overall economic and international market distortions, and trade policy tensions. Positive effects generated by well-designed
public policy measures should, in turn, be realized in enhanced competitiveness of a certain industry facing open competition.

Given that public sector activities and general economic policy are also
considered to be of significance with respect to industry competitiveness,
policy measures could be of the following nature (Romppanen and
Leppänen, 1995, p. 185):

- removal of trade barriers (the focus of a growth policy)
- reduction of real interest rates by the improvement of economic stability and economic policy credibility
- increase in human capital through public sector investments in education, research and health care, etc.
- stability of the financial system secured by sensible institutions and leg-
  islative frameworks

- flexibility of the labour market
- taxation incentives, income transfers and services that promote efficiency and entrepreneurship
- competitiveness of the physical infrastructure.

The objectives of economic growth, efficient allocation of resources and improvement in economic flexibility can be realized only with the aid of indirect public sector incentives (Romppanen and Leppänen, 1995, pp. 185–7). The tools are:

- legislation, i.e. the clarification of the economy's own internal rules
- the allocation of public expenditures and taxation
- production and dissemination of information primarily in the form of education and research.

In addition, an important task of public policy is the correction of market failures. Markets are not able to take responsibility for education, research and innovative activities required for infrastructure creation (especially modern information superhighways) and provision of attractive factor conditions. Market forces are also unable to deal with environmental issues, which are becoming more and more important in all sectors.

Essential government policies affecting the overall environment, functioning, and competitiveness of the food industry are agricultural, industrial, competition, trade, investment and R&D policies. The objective of policy is not to maintain the status quo by inefficient intervention such as subsidies and protection, but to provide the environment and incentives for an industry to improve its own ability to compete in an open market.

In the EU, however, the CAP still strongly affects the functioning of food chains in all member states. Indeed, policy and government interventions play an important role in agriculture globally. Various measures of price and income support have been applied to achieve the farm income goal. The results have often been poor:

- the farm income objective has not been reached
- inappropriate means of support and regulation have been used, structural development has been hindered
- negative externalities have been created
- excessive costs have been incurred by citizens through high consumer prices and taxpayer costs.

Government interventions and subsidies to correct the perceived market failure of too low and volatile agricultural prices and incomes have often turned out as a policy failure. In this respect agricultural policy has not functioned efficiently as a public policy.

In the case of industrial policy, in an increasing majority of countries the role of government is rather seen as influencing only indirectly the business environment in which industrial enterprises can prosper (Gassmann, 1996).

The archetypical form of industrial policy was characterized by direct government intervention in the form of, for example, state ownership of industry, direct subsidies to mature or declining sectors and to underdeveloped regions within countries, and border protection of domestic markets. Today the approach is more and more towards horizontal measures affecting all sectors indifferently, such as subsidies to promote SMEs. Paradoxically, as Gassman (1996) points out, this development towards a more modest role for the state may involve a wider range of government institutions in order effectively to coordinate diverse ingredients of industrial competitiveness policy.

The food sector has two characteristics that distinguish it from other industries (e.g. Volk et al., 1996). First, it is controlled by the CAP, which sets the institutional framework for the entire industry, not only for farming. Secondly, the food sector is a mature industry, producing consumer goods with slow growth potential. Industrial policy has no major interest in this kind of static industry. Governments are usually interested in maintaining, preferably increasing, competitive, profitable economic activities within their country. Eventually, the basic question for public policy in general, and industrial policy in particular, is how to guarantee that the country is an attractive location for firms of high international competitiveness. Although countries do not compete directly, only firms do, indirectly countries compete more and more by improving the quality of their business environment and providing favourable factor conditions.

The theoretical basis of our integrated approach of looking at public policy instruments and industry competitiveness together is set in the model of the Porter diamond (the upper left corner of Figure 8.1). Instead of focusing on one policy only, here we look at several possible public policies affecting the food economy, its strategies and competitiveness. In addition to understanding the nature of the determinants of competitiveness, we should also understand the dynamics and the nature of the rivalry and the interaction between the different parts of the model (see Chapter 1). Instead of oversimplified and static models such as the structure–conduct–performance (SCP) model (Bain, 1968; Scherer and Ross, 1990; Viaene and Gellynck, 1995 for a recent application of the SCP model to the European food sector), we have a more dynamic and complex model in the Porter diamond. Although public policy is also indirectly present in the SCP model, we want to emphasize its role by establishing more concrete and direct linkages from public policy to demand and factor conditions, to related and supporting industries, and to strategies, structures and rivalry of firms in the sector. Our public policy framework includes both general legal and administrative actions of government, which aim at improving the quality of the business environment. The ultimate aim is to identify effects of policy instruments on individual food firms' resources and competitive position in the SEM.

An applicable framework for the evaluation of the effects of public policy measures is provided by Porter's approach. It shows that competitiveness is

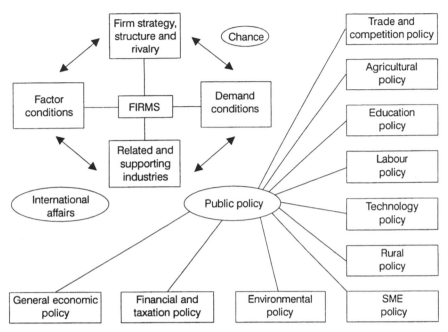

Figure 8.1 Public policy and business environment (adapted from Hernesniemi *et al.*, 1995).

created through an interaction between different industries, firms, and the public sector (Figure 8.1). Public policy can promote the gaining of competitive advantages, but it cannot create them.

- Through **competition policy** the public sector creates the environment in which the firms build their competition strategies. Heavy competition creates competitive advantages, lack of competition weakens firms' international competitive position (e.g. in the Finnish food industry).
- **Education and technology policies** improve factor conditions, which are crucial to economic growth and international competition, including competition for foreign investment.
- Demand conditions of firms are strongly affected by, for example, **taxation and trade**, and more and more by **environmental policy**. In addition, development of related and supporting industries and creation of firm networks can be enhanced by, for example, created factor conditions or competition policy. In the jargon, the diamond of competitive advantage shows the mechanism through which the effects are transmitted (Figure 8.1).

The dependence of agriculture (farms) and agribusiness (input suppliers and food manufacturers) on public policy is unambiguous in the industrialized world, one of the best (worst) examples being the CAP of the EU. However, the theoretical framework for analysing linkages of public policies and firm strategies is somewhat ambiguous, and varies according to the policy under scrutiny. Hence, our aim is to cover a complex and dynamic environment of numerous interdependencies by the application of the Porter approach with a special emphasis on the role of the public sector and public policies.

It should be acknowledged that although we have chosen a Porter approach for our analysis, other methodologies may be appropriate under other circumstances, e.g. strategic trade policy theory to assess whether governments would benefit from subsidizing firms which possess market power on world markets (e.g. Eaton and Grossman, 1986; McNally, 1995), or an analysis of the welfare allocation effects of a major restructuring of an industry. The aim could be at finding out whether the increased cost efficiency (for example through merger-induced consolidation in an oligopolistic market) is sufficient to offset the allocative inefficiency resulting from greater market power (e.g. Azzam and Schroeter, 1995; Marttila, 1996).

8.3.2 Competitive advantage of firms: two alternative approaches

A competitive strategy specifies how a business should compete in a given industry or product-market arena in order to develop sustainable positional advantages. Competitive advantage can be defined as a unique position which a firm develops *vis-à-vis* its competitors (Day, 1990). There are two complementary approaches to studying competitive advantage. One model is based on Porter's (1980) generic strategy typology, which originally relied on industrial organization (IO) theory, and later on Porter's diamond model. The second approach draws on resource-based theory (Conner, 1991; Mahoney and Pandian, 1992). Furthermore, a competitor-centred or a customer-centred view can be adopted in the study of competitive advantage. Within the scope of this research, our focus is on the competitor-centred view.[1] Next we describe the two approaches explaining competitive advantage.

Positional advantages. Drawing on industrial organization theory, Porter (1980, 1985) combines the elements of competitive strategy and performance paradigm in a model of the structural analysis of industries which is based on the evaluation of five competitive forces that determine the nature of competition in an industry (i.e. the conduct of its participants) and its underlying profit potential. The unit of analysis is principally the industry[2] but, as Porter (1985) argues, the model can also be applied at the level of strategic groups

1. According to Day and Wensley (1988, pp. 2–5), customer-focused approaches to competitive advantage centre on identifying specific key 'success factors' that a firm should deal with to succeed in the market. Major emphasis is on the consumer perceptions of product attributes, satisfaction, loyalty, etc., and relatively little attention is paid to the competitors' actions. In reality, however, the customer-centred and competitor-centred approaches are interrelated, and they should be used together.
2. Referring to Porter's diamond model and its modification (Figure 8.1), this view is related to firm strategy, structure and rivalry, but in our case characterizing domestic rivalry.

within the industry. The firms in a strategic group will tend to have similar market shares and be affected by and respond similarly to external events or competitive moves in the industry because of their similar strategies. The strategic group is therefore an intermediate frame of reference between the level of the industry on the one hand, and that of the individual firm on the other. Porter (1980) also introduces a typology or a conceptual classification of three generic competitive strategies through which firms 'beat the competition' to gain positional advantages. Cost leadership aims at the lowest cost position in the industry. The development of competitive advantage may then include the pursuit of economies of scale, experience and learning effects, and tight cost control in production. In differentiation strategy, a firm strives to be unique along some dimensions that the buyers in numerous market segments perceive as important, and by positioning the firm to uniquely meet those needs. Differentiation may be achieved through a strong brand name and company image, product and service quality, new technology, a particular distribution system or a broad product line. A focus or protected niche strategy concentrates resources on building and defending deep positions in relatively narrow market segments. Focusing on one or two segments in an industry may either provide a cost advantage or allow differentiation.

Most of the empirical studies which have adopted the generic strategy approach explore the strategy-performance relationships at strategic group level (e.g. Dess and Davis, 1984; Thomas and Venkatraman, 1988). Some authors argue, however, that the typology is oversimplified, and it may lose its explanatory power in an examination of the strategies and performance of individual firms (Hill, 1988; Murray, 1988; Brownlie and Moutinho, 1990/91). More recently, Porter (1991, p. 101) stresses that one should make a distinction between the position of advantage reflecting a sustainable competitive advantage and the sources of advantage. According to him,

> an attractive relative position results from possessing sustainable competitive advantage (i.e. defined as earlier) within some scope, and several positions can be attractive in absolute terms. Scope encompasses a number of dimensions including product-market domain, the geographic locations in which the firm competes, the degree of vertical integration, and the extent of related businesses. The sources of competitive advantage, in turn, center around activities ('what firms do') which can be described in terms of the value chain.

To some extent this view parallels the resource-based view of competitive advantage.[3]

Sources of competitive advantage.   The resource-based model emphasizes the distinctiveness of a firm's resources as the sources of competitive advantage

---

3. Porter (1991, p. 108), however, is critical of extant versions of resource-based theory: 'At worst, the resource-based view is circular'. For him, this view is inward looking. Instead, selecting the right industry and generic strategy within an industry is key because industry is the most significant predictor of firm performance (Montgomery and Porter, 1991, p. xiv).

in exploiting market opportunities (Selznick, 1957; Penrose, 1959; Wernerfelt, 1984; Day and Wensley, 1988; Conner, 1991). Although the unit of analysis is the individual firm, analysis of the environment is critical since several forces such as macro-environmental, competititive and market changes may alter the importance of resources to the firm (Penrose, 1959). The analysis of firm-specific resources extends quite naturally to international business competition and co-operation.

Wernerfelt (1984, p. 172) broadly defines a **resource** as 'anything which could be thought of as a strength or weakness of a given firm'. Resources are the tangible and intangible assets available to the firm that enable it to produce efficiently and/or effectively market an offering that has value for some market segment or segments. **Value** refers to the sum total of all benefits that consumers perceive they will receive if they accept the market offering (Hunt and Morgan, 1995, p. 6). As to the resources, transforming inputs into outputs requires the deployment of tangible resources such as a highly automated production equipment, broad distribution coverage, working capital, advertising and product development expenditures. For example, technological superiority and marketing resources provide a firm with the capability to generate new products/processes faster than competitors. On the other hand, intangible assets can be illustrated by a variety of characteristics such as brand images, reputation, the relationships to suppliers and buyers, and a good knowledge of customer needs (Day and Wensley, 1988). Intangible resources can accumulate over time, provided that the environment remains relatively stable. For instance, a firm's present reputation is based on its previous good relationships with customers and their perceptions of 'value' and quality products/services. On the other hand, a strong brand image is a result of heavy past advertising, which can then make a current rate of spending more effective. Furthermore, a firm's core competences[4] (Prahalad and Hamel, 1990) are intangible, higher-order resources that enable it to perform the activities in specific ways, better perhaps than its competitors. Different types of assets are also more or less durable, and some may be easier or harder to imitate than others. Intangible resources – often not directly associated with a product or service – take time to cultivate and replicate.

According to Barney (1991), the multitude of potential resources can also be categorized as

- **financial** (e.g. cash reserves, access to financial markets)
- **human** (e.g. skills)
- **organizational** (e.g. competencies, culture, policies)

---

4. Some studies make an explicit distinction between resources, competences and capabilities (e.g. Mahoney and Pandian, 1992). Resources refer more to possession (having) while competences and capabilities to use (doing). Obviously, the competitive advantage does not arise from the possession of resources *per se*, but also from a capability to make better use of resources (Penrose, 1959; Hunt and Morgan, 1995).

- **legal** (e.g. licences, trademarks)
- **informational** (e.g. knowledge resulting from consumer and competitor intelligence)
- **relational** (e.g. relationships with suppliers and buyers).

When a firm has unique resources that are rare among competitors, the potential for producing a comparative advantage for that firm exists. A comparative advantage in resources, then, can translate into a position of competitive advantage in the marketplace and superior financial performance – depending on whether those resources are also valuable, imperfectly imitable, and/or strategically substitutable.

In our study, competitive advantage is defined in terms of tangible and intangible resources that here refer to a set of special activities that the firm performs particularly well or differently relative to rival firms within a similar product-market domain.

8.3.3 Marketing-related resources

The marketing function, especially the marketing mix, plays a dominant role in the context of leveraging the tangible and intangible resources into sustainable positional advantages in the target market of a firm. On one hand, the uniqueness of the mix depends on how much a firm allocates resources on marketing, on the other hand on the positioning of the product in the target market(s). Our focus is on the allocation of resources.[5] In the following section, the key marketing variables are described.

Product mix. Decisions concerning the product mix of a firm reflect the breath and depth of the product range, and the selection of an appropriate branding strategy. Product decisions can also involve the prices of products to be offered. The breath of product range describes scope commitment decisions; product variety is itself a dimension of market performance (Cravens, 1987; Day, 1990). According to Cravens (1987), there are three major brand strategy options available for manufacturers:

- no brand identity
- corporate/product branding
- private branding strategy or the manufacturing of distributor brands.

In many product groups vertical competition from distributor brands provides a sharper challenge to the brands of several leading manufacturers than do those of secondary manufacturers. In the 1990s, retail chains have

---

5. Concerning marketing expenditures that are caused by the marketing mix (as being important in determining business success and profitability), the majority of empirical research has been based upon the PIMS database, which contains primarily large firms. However, these studies reveal only a few consistent relationships, allowing little generalization (see e.g. Day, 1990).

increasingly introduced their own private brands in many lines in order to differentiate their stores (see Chapter 2). Retailer margins are higher on private-brand items than on comparable products carrying a manufacturer brand (Reibstein, 1995). The opportunity for better profit margins arises through growing bargaining power resulting from the increasing levels of concentration and increased centralization of purchasing. Retailers may seek suppliers who can provide products to the positioning of the store (e.g. price, quality and innovation) rather than, as in the case of manufacturer brands, taking largely predetermined products. The contract manufacturing of distributor brands can offer lower risk and better cash flow than is possible with the costs of a national launch and gaining national distribution. Those suppliers may reap greater rewards in terms of trade customer loyalty than is the case for manufacturers' brands. Retailers promote private brands by giving such lines more shelf space and by training staff to recommend them. The distributors may also entice manufacturers of branded goods to allocate marginal production capacity for producing distributor brands (Shaw et al., 1992). However, the loss of control in switching to the manufacturing of private brands is considerable, and the only remaining competitive parameters are costs and consistent quality; also, smaller manufacturers are not able to satisfy the requirements of large volume sales. In the study, branding strategy is described in terms of corporate branding versus private branding strategy, i.e. whether or not a firm manufactures distributor brands.

The breadth of product range, branding strategy, and the characteristics of the distribution channel network often set the framework for pricing of products. In addition, product quality, end users served, and intermediate functions, including the form and structure of discounts for volumes, terms of payment, and trade-in allowances all tend to affect pricing strategy (Hardy and Magrath, 1988). Often, manufacturers may pursue low prices in pursuit of market share, or they may opt for parity prices as part of an effort to maintain their share of the market, or they may choose premium prices as part of a skimming strategy. Primarily, manufacturers must establish pricing structures that encourage their distributors to perform the desired functions of stocking and merchandizing the product lines. Although differentiation through non-price factors has become a profitable strategy for many larger retailers, price is still an important criterion in the retailer's purchasing decisions from different suppliers (e.g. Shaw et al., 1992). For example, there is evidence to indicate that lower-priced campaign products are a widely used method in the Finnish food chain (Hyvönen, 1993). In the study, pricing as a marketing mix variable reflects the extent to which a low price strategy has been adopted by manufactures.

Product development. Through product development activities firms seek opportunities that permit the use or extension of their special technical, production, or marketing skills, or the employment of resource base. The underlying strategic purpose should always be either to help create and maintain a competitive advantage or to reduce the advantages of a competitor. Categories of new products usually range from completely new products to modifications of existing products and new targeting and positioning of existing products (Cravens, 1987). The probability that expenditure on product development will result in sustainable differentiation relative to competitors diminishes over the product/industry life cycle. In mature markets growth is low, the market structure can be concentrated, brand loyalty is well established, and processes are well developed. Other bases that are difficult to imitate must then be found, e.g. technological change that leads to new processes and better quality products. Moreover, product development is strongly dependent on other marketing mix variables (e.g. breadth of product range, advertising) for its successful implementation. In convenience goods industries, the innovating firm will not generally actualize the benefits of introducing new products unless the market introduction is accompanied by heavy advertising and sales promotion.

In contrast to differentiation,[6] product development is a variable that managers have control over, reflecting resource commitments. In the study, product development is described in terms of the number of new products introduced into the market, product development expenditures, and their change during the last 5 years.

Advertising and trade marketing support.   Firms usually use two basic communication strategies reflecting advertising (Stern and El-Ansary, 1992).

- The **pull strategy** is aimed at the ultimate consumer or user rather than the intermediary and directly attempts to create demand for the manufacturer's goods so as to persuade distributors to place orders and stock them.
- The **push strategy** aims at intermediaries in the channel of distribution such as wholesalers and retailers and their industrial equivalents. The promotional objective is to encourage the stocking and display of products so that ultimate consumers/users receive the maximum exposure to them (Olver and Harris, 1989). Push strategy requires effort and support from the manufacturer's sales force, and promotional programmes directed specifically towards the distributor.

---

6. In industrial organization literature, product differentiation reflects a structural indicator of entry barriers, and it has been measured based on industry-level advertising expenditures (Porter, 1980; Day, 1990). In order for a product differentiation strategy to be viable, a firm should build and sustain noticeable differences in its product offerings or in brand image, packaging, and associated services. As an industry matures, competitors' product offerings tend to converge towards those product offerings most preferred by customers, and products do not differ much from each other physically (Porter, 1985). Instead, there may be perceived degrees of differentiation that are based on advertising, packaging or subjectively experienced characteristics. It is therefore extremely difficult to develop valid measures of the 'degrees' of differentiation.

One category of trade marketing support can be characterized as payment for services rendered by wholesalers or retailers. These payments are discounts given to wholesalers and/or retailers for displaying, and/or advertising a manufacturer's product over some period of time. Such programmes as instore displays, display allowances, and co-operative advertising are examples of this category. The second category consists of buying allowances and trade deals. For example, a quantity discount system compensates a channel member for buying in large lot sizes, such as the discounts available in volume purchase agreements. A functional discount system compensates a channel member for its participation in transactional flows. Direct financial support is a form of functional discount. Manufacturers give such discounts to ensure distributor support in the hope that the discounts will be passed on to end users in order to stimulate additional demand. Compensation in distribution should be given on the basis of the degree of each channel member's participation in the marketing flows. However, it is rarely employed. Because retailers differ in the size of their physical establishment and the volume of their sales, allowances obviously cannot be made available to all retail buyers on the same absolute basis. The same holds for supplier firms regarding the contribution of retailers to the promotion of suppliers' products. Hence, it is common that allowances are also made for services not actually rendered.

In the study, a firm's advertising expenditures and their relative change reflect the intensity of its pull strategy, and a firm's trade marketing expenditures and their relative change reflect the intensity of its push strategy.

Marketing organization and firm-specific variables. We propose that the size of marketing organization is an important resource commitment variable affecting the competitive advantage of firms, e.g. ability to develop differentiation. We also introduce two contextual variables: the size and age of the firm.

8.3.4 Organizational performance

The importance of the performance concept (and the broader area, organizational effectiveness) is widely recognized, but the treatment of performance in research settings is perhaps one of the thorniest issues confronting strategy research today. According to Venkatraman and Ramanujam (1986), most prior studies on strategy have described performance in terms of profitability, either alone or together with other performance indicators. Because profitability is influenced by actions taken in many previous time frames, it is unlikely to be a complete reflection of current advantage. The interpretation of profitability is complicated further by limitations in the prevailing modes of valuation:

> The cost-based approaches that underlie most accounting results are fundamentally different from approaches that estimate financial value from the stream of future

benefits. *Accounting conventions oriented to allocating historic and current costs* to satisfy tax requirements are *ill-suited to the valuation of the sources of advantage.* The consequences of the accounting mindset are most evident in the treatment of intangibles. Goodwill becomes an arithmetic necessity rather than a genuine commercial asset with future value. Similarly, investments in the skill and knowledge base are treated as current overhead, without consideration of their contribution to long-run performance (Day and Wensley, 1988, p. 5, italics added).

Given that performance is not a unitary concept, considering only profitability as a measure of performance is inadequate. Access to accounting data on privately held firms can also be severely restricted. In this study, organizational performance is described in terms of subjective measures that reflect profitability, efficiency, and effectiveness criteria. Efficiency indicates input–output ratios internal to the firm while effectiveness reflects 'how well an organization relates to its environment', for instance by successfully expanding its product-market scope.

8.4   Research methodology

8.4.1   Research questions for empirical analysis

In identifying the types of competitive advantage or strategy patterns, it is possible to choose among two classificatory schemes (Miller and Friesen, 1984). The first is called a **conceptual classification** or a **typology** in which the patterns of strategy are deductively derived. A well-known typology is that of Porter (1980). The linkage between *a priori* specified analytical dimensions and their corresponding measures has, however, generally been weak (cf. Hill, 1988; Murray, 1988). The second scheme is called an **empirical classification** or a **taxonomy**. It derives strategy patterns *a posteriori* based on the analysis of field data. In this study, the latter approach is applied. Consequently, two basic research questions are addressed:

- Are there consistent patterns of strategy developed by food manufacturing firms, i.e. what are the most important dimensions of the sources of competitive advantage?
- Can firms be meaningfully classified on the basis of strategy patterns? Do these clusters of firms differ systematically in terms of key marketing variables, organizational characteristics, and organizational performance?

8.4.2   Field study and sample

The empirical study is cross-sectional and was conducted in a field setting in the Finnish food industry. Firms operating in the meat processing, soft drinks and milk processing industries were selected as the most appropriate data source. The total value added contributed by these three sub-industries is about 45%, suggesting that they are relevant sectors in our food industry. The initial list of companies was drawn from those listed in the published statistics of the Ministry of Agriculture and Forestry and Finnish Food and Drink Industries. In total, 88 operating firms were identified, and all of them were selected for empirical exploration. This setting permits the examination of relevant questions applicable to diverse firms while controlling for circumstances that might otherwise vary greatly across industries. There are firms of differing sizes which operate in different market segments. This should ensure enough variability to study strategic behaviour.

A semi-structured questionnaire was mailed either to the chief executive officer (CEO), chief marketing executive (CME), or the owner of the company. These top managers are invariably the persons most involved in the development and implementation of business-level strategies. While strategy is an organizational-level, not an individual-level, concept, the use of multiple informants could violate some fundamental assumptions of strategy. The two mailings and personal contacts resulted in a sample of 65 usable questionnaires, a response rate of 73.4% (65/88).

A comparison of early-responding firms and late-responding firms (those that responded after a follow-up letter was sent) showed that these groups do not differ in terms of years of business, number of employees or firm performance. Concerning the industry categories of firms, 58 producers represented the meat-processing industry, 4 the soft drinks industry, and 3 the dairy industry. The firms averaged 204 employees and 32 years in operation. Small firms employing less than 10 persons accounted for one-third of the sample. A total of 8 firms had operated for less than 4 years.

8.4.3 Measurement of variables

The survey instrument was developed according to the general approach recommended by Nunnally (1978). Since competitive strategy issues are proposed to be unique to a particular setting, industry-specific conditions as well as theoretical aspects should be carefully taken into account in designing measures. Consequently, several iterations of the research instrument were made prior to an actual field test. First, input for the development of measures came from prior literature and documentary data concerning the food industry. Second, the relevance of the items was ascertained through the use of extensive interviews with the CEOs and CMEs in six firms, which totalled 12 interviews. Third, a panel of academic experts provided recommendations for developing measures. The interviews led to several improvements in both the wording and the composition of lists of variables.

Competitive advantage. Competitive advantage was defined broadly, reflecting various tangible and intangible resources for the purpose of providing a general profile of the concept. As such, production, purchasing, marketing, distribution, and finance variables are all represented in the competence scale which consists of 19 items (see Table 8.3). Some items were adapted from earlier studies (Dess and Davis, 1984; Robinson and Pearce, 1988). Respondents were asked to indicate the degree to which their firm emphasized each of the listed 'success' factors or methods of competing. Seven-point scales with values ranging from 1 ('not at all important') to 7 ('extremely important') were used. The instructions to the respondents also stressed that they should use their major competitors as a frame of reference, and that they should selectively emphasize particular competitive methods.

Marketing-related resources.   There are five indicators for marketing-related resources. Taking into account the limited scope of the study, only one variable measured each indicator. New product emphasis was operationalized in terms of the number of new products launched during the previous 2 years. The variable 'the proportion of lower-priced campaign products in the product range' indicates the extent to which a low-price strategy was adopted by food manufacturers. Both advertising and trade marketing expenditures were expressed as a percentage of sales. The size of the marketing organization, which reflects marketing capabilities and the related expenditures, was operationalized in terms of the number of employees working at marketing tasks. Two contextual variables were also introduced into the analysis, the size and age of the firm.

Organizational performance.   There are three indicators of organizational performance, reflecting profitability, effectiveness and efficiency. Following the recommendations of Dess and Robinson (1988), we developed a self-reported scale on profitability in the following way. Managers indicated to what extent their firm had realized its performance objectives concerning the following five criteria: sales growth rate, gross margin, operating margin, net profits from operations, and return on shareholder equity. The response options for the scale ranged from 1 ('highly dissatisfied') to 7 ('highly satisfied with the realized result'). The unidimensionality or internal consistency of the scale was measured by Cronbach's alpha. Since the scale had a high internal consistency ($\alpha = 0.87$), the five items were combined into a summed scale measuring overall financial performance. The sales growth rate over the last 5 years measured effectiveness, and efficiency was defined as sales turnover/the number of employees working in food processing.

8.4.4 Aggregation of firms: evaluation of industry effects

Previous studies provide some evidence that business level strategies and their performance effects may vary by industry (Venkatraman and Ramanujaman, 1986). Consequently, prior to aggregating the firms independent of their industry classification, a comparison was made of the competence and performance variables among the firms grouped by the three industries. ANOVA results indicated that there were significant differences in only 2 of the 19 competence variables ($p < 0.05$ and $p < 0.10$, respectively). To identify the industry category sources for this variation, the Scheffe multiple comparison test was performed. It revealed no paired comparisons significant at the 0.05 level. No significant differences were found among performance variables by industry, either ($p < 0.05$). Thus we may conclude that the industry effects are minimal in this study, and therefore it was appropriate to aggregate the responses.

8.5   Analysis of field study results: how do Finnish food manufacturing firms compete?

8.5.1   Identification of competitive strategy patterns

In order to identify the most important dimensions of the sources of competitive advantage or strategy patterns, the variables were analysed through factor analysis. An orthogonal rotation (VARIMAX) of the initial principal components factor matrix yielded five factors. A combination of minimum eigenvalue criterion and scree test were used to determine the number of factors. The results are shown in Table 8.3. All 19 original variables exhibited factor loadings greater than or equal to [0.42] on at least one factor. These loadings may be considered as a conservative criterion (Kim and Mueller, 1978). Because of the potential instability of factor scores with 65 firms and 19 strategy variables, the factor analysis was performed five times on $(n - 1)$ cases in order to test whether a changing of sample composition would alter the factor loadings. The analysis results were the same or similar in all runs. Altogether five items loaded on two factors. However, these variables were not deleted because a firm's competitive advantage can consist of a combination of tangible and intangible resources (Day and Wensley, 1988). For instance, the firm may simultaneously have marketing-related and manufacturing-related strengths.

   Marketing differentiators (factor I) compete with a broad product range typically involving speciality products and with a strong emphasis on product development and new technology. Direct advertising, brand marketing and a skilled sales force are also important dimensions of marketing differentiation. Dess and Davis (1980) have identified a similar type of strategy as the basis of competitive positioning. **Distributor orientation** (factor II) suggests an especially strong commitment to 'push marketing' and a large marketing and sales organization necessary for its implementation. These firms emphasize serving specific market segments; for example, they manufacture distributor brands for the integrated retail chains. In earlier studies, a competitive strategy called 'brand and channel influence' has been found (Robinson and Pearce, 1988). Image and product development (factor III) shows high loadings on good corporate image, tight quality and cost control, and product development and innovation. This combination suggests a strategy based on efficiently producing a narrow line of niche products. Reputation and high-quality products may be essential dimensions of focused strategy (Porter, 1980). Production and supplier orientation (factor IV) is characterized primarily by cost advantages based on economies of scale. This focus on large-scale manufacturing is combined with a strong emphasis on the quality and availability of raw materials, the control of distributors, and operating efficiency including new technology. Such an orientation is consistent with previous studies that identified 'an overall low-cost strategy' (Dess and Davis, 1984) and 'emphasis on efficiency in operations' (Robinson and Pearce, 1988). Porter (1980) also contends that 'achieving a low cost position often requires a high relative market share or other advantages, such as favorable access to raw materials' (p. 36). Factor V, 'low-priced products with no brand identity', was of minor importance in the factor structure. These firms do not manufacture well-known brands (negative loading); their strength is in price competition.

8.5.2 A taxonomy of competitive advantages

In the second phase, the factor scores obtained from the above analysis (Table 8.3) were utilized as the input variables to classify the firms. Ward's hierarchical centroid method based on squared Euclidean distances was used to form clusters. This method is considered to be one of the most accurate hierarchical clustering techniques. Ward's method produces a grouping of relatively homogeneous groups of firms which have maximum between-group variance and minimum within-group variance (Punj and Stewart, 1983; see also Everitt, 1980). There is no absolute criterion for selection of the number of clusters. The decision rule for formation of homogeneous clusters was based on

- minimum error sum of squares
- inspection of the dendrograms displaying the contents at each stage
- examination of the cluster contents in terms of the underlying concepts of interest.

A six-cluster solution was found to maximize the distances between cluster means across the five factor patterns. A four- and a five-cluster solution were also examined. The resulting clusters, however, were not meaningful. To enhance the interpretation of each cluster, cluster means (mean value of factor scores) for the six-cluster solution as well as the number of firms in each cluster are reported in Table 8.4.

Cluster 1 has its highest mean value of factor scores on a distributor-oriented strategy; it also has a (low) positive score for the strategy empha-

274 COMPETITIVENESS IN THE FOOD INDUSTRY

Table 8.3 Factor structure of competitive advantages (values $\geq$ 0.42)

	I	II	III	IV	V	H²
			Factor loadings			
Good corporate image			0.81			0.69
Product quality control			0.70			0.64
Continuing concern for lowest cost per unit			0.56			0.65
Competitive pricing					0.85	0.77
Economies of scale based on mass production				0.42	0.43	0.74
Building brand identification					−0.53	0.69
Finance and operating efficiency					0.68	0.74
Major effort to ensure the availability of raw materials				0.87		0.84
Emphasis on trade marketing		0.70				0.68
Control of distribution channels				0.42		0.64
Emphasis on production processes and new technology	0.46			0.51		0.71
Strong marketing and sales organization	0.46	0.72				0.79
Continuing product development and innovation		0.54		0.48		0.75
Focus on specific market segments		0.80				0.71
Broad range of products	0.87					0.80
Capability to manufacture speciality food products	0.55					0.52
Depth of product range, large number of items	0.79					0.76
Advertising expenditures above the industry average	0.67					0.69
Manufacturing of distributor brands	0.47	0.45				0.55

I, marketing differentiation; II, distributor orientation; III, image and product development; IV, production and supplier orientation; V, low-priced non-branded products.

	I	II	III	IV	V
Eigenvalue	7.01	2.10	1.76	1.40	1.07
Percentage of cumulative variance accounted for	36.9	48.0	57.3	64.7	70.3

sizing price competitiveness with no brand identity. Marketing differentiation and image and product development (high negative values) are not important for these firms. Cluster 2 does not emphasize any particular pattern of strategy because it has positive scores on all strategies. Therefore, it may be comprised of firms that are 'stuck in the middle' (Porter, 1980). Cluster 3 is clearly oriented towards marketing differentiation, lacking those cost-based advantages that relate to the sources of supply (a high negative score on production and supplier orientation). Cluster 4 emphasizes differentiation that is primarily based on image and product development. There are also firms focusing on marketing differentiation. This group has a high negative score for a production-oriented strategy and for a low-price strategy with no brand identity. Cluster 5 is primarily characterized by a production and supplier orientation. The group does not emphasize distributor orientation or a low-price strategy. The firms in cluster 6 consider a distribution-oriented strategy most important, also emphasizing differentiation based on image and product development. The cluster has a high negative score for production and supplier orientation as well as for marketing differentiation. Taken in combination, the pattern of mean scores that emerged from the cluster analysis shows relatively high and positive scores on several alternative strategy types. This may indicate the possibility of emphasis on more than one strategy within the groups of firms.

Next, differences between clusters on strategic variables not utilized as bases of classification are described. In particular, marketing-related resources and organizational characteristics were examined using ANOVA. Duncan's multiple range test was used in order to determine which group means were different from each other. Table 8.5 shows the strategic profiles of the six groups of firms along with F-statistics and Duncan's tests. Based on relative emphasis on differing types of advantages as shown earlier (Table 8.3), descriptive labels are also provided.

- **Cluster 1**: distributor-oriented, low-price strategists (16% of sample). This cluster consists of larger food manufacturing firms that operate on a regional and national scale. These firms had launched an average of 11 new products during the previous 2 years. A relatively high proportion of lower-priced campaign products (an average of 49% of product range) is an indication of a strong distributor orientation. The marketing organization averages 17 staff members.
- **Cluster 2**: utilizers of differentiation- and cost-based advantages (22% of sample). This consists of smaller and medium-sized firms which have no clear strategic orientation. During a 2 year period there were 13 new product introductions, but a low-price strategy is not widely adopted in

Table 8.4 Six cluster solution: cluster means

Cluster	Factor I: Marketing differentiation	Factor II: Distributor orientation	Factor III: Image/ product development	Factor IV: Production/ supplier orientation	Factor V: Low-priced non-branded products
1 ($n = 8$)	–0.8809	**0.1331**	**–0.8250**	–0.1059	0.0808
2 ($n = 11$)	0.1200	**0.5524**	**0.4513**	0.3793	**0.6569**
3 ($n = 7$)	**1.5891**	0.2694	–0.7863	**–1.3979**	0.5822
4 ($n = 12$)	**0.5909**	–0.9744	**1.9811**	–2.2783	**–1.5661**
5 ($n = 4$)	0.1183	**–0.7797**	–0.0769	**0.3422**	–0.6215
6 ($n = 8$)	–1.0690	**1.3549**	**0.5157**	–0.9163	**–1.4719**

this group of firms. This 'stuck in the middle' cluster has the lowest advertising and trade marketing expenditures. The marketing organization consists on an average of eight staff members.

- **Cluster 3**: marketing differentiators lacking cost-based advantages (14% of sample) This cluster consists of large firms. The firms compete with a broad product range, having launched on average 50 new products over the past 2 years. The proportion of campaign products is large, averaging 65% of product range. Consequently, a low-price strategy is commonly adopted by the firms. Cluster 3 contributes heavily to trade marketing. Advertising expenditures are the highest, and the marketing organization the largest, averaging 54 employees.
- **Cluster 4**: innovating differentiators lacking cost-based advantages (24% of sample). This cluster contains medium-sized and smaller firms. The group had launched an average of 15 product innovations during the previous 2 years. While the proportion of lower-priced campaign products accounts for an average of 40% of the product range, to some extent the group emphasizes a low-price strategy. Advertising and trade marketing expenditures are relatively low. On the other hand, the marketing organization is relatively large, averaging 18 staff members.
- **Cluster 5**: production- and cost-oriented strategists lacking marketing competence (8% of sample). This cluster includes small firms with a very small marketing organization, averaging only 1.2 persons. The group had introduced an average of 12 new products during the previous 2 years. A low-price strategy is not common in this group.
- **Cluster 6**: distributor-oriented image and product developers (16% of sample). This cluster consists of very large firms that have operated in the business for a long time; firm age averages 53 years. The group most strongly emphasizes a push strategy oriented to their distributors, which can be seen from the high proportion of trade marketing expenditures. A relatively high proportion of lower-priced campaign products demonstrates a focus on competing on the basis of price in channels of distribution. In the group, the size of the marketing organization is the second largest, averaging 45 staff members.

The data revealed that some firms in clusters 3, 4, and 6 manufactured distributor brands (the results are not reported here). This explains, in part, their focus on distributor orientation. Differences in marketing variables by cluster provide some confirmation of the validity of the cluster solutions. As Table 8.5 shows, ANOVA reveals significant group differences on four marketing variables, and three of them were marginally significant ($p < 0.10$). Duncan's comparisons that show the sources of significant differences are also reported in the table.

Table 8.5 A taxonomy of competitive advantages in the Finnish food industry

	Cluster/cluster title						
	1	2	3	4	5	6	F-stat
Number of new products launched	10.7	13.2	50.0	14.9	11.7	24.4	2.400*
Proportion of lower-priced campaign products	48.6	12.4	65.0	40.0	10.3	56.7	4.417***
Advertising expenditures (%) of sales turnover	1.7	0.7	2.6	0.8	0.9	2.0	4.173***
Trade marketing support (%) of sales turnover	1.4	0.5	4.3	1.4	1.2	6.1	2.143 ^
Size of marketing and sales organization	17.5	7.9	53.6	17.6	1.2	44.7	3.708**
Years of operation	26.6	29.0	46.4	22.8	48.5	52.6	2.176 ^
Sales turnover (Million FIM)	236.2	58.0	291.3	161.1	32.6	397.1	1.970 ^
% of sample	16	22	14	24	8	16	

1, distributor oriented, low-price strategists;
2, utilizers of differentiation- and cost-based advantages;
3, marketing differentiators lacking cost-based advantages;
4, innovating differentiators lacking cost-based advantages;
5, production- and cost-oriented strategists lacking trade marketing competence;
6, distributor-oriented image and product developers.
*** $p < 0.001$; ** $p < 0.01$; * $p < 0.05$; ^ $p < 0.10$.
Duncan's comparison ($p < 0.05$):
    Number of new products CL3 > CL1, CL5, CL2, CL4;
    Proportion of campaign products CL3, CL6, CL1, CL4 > CL5, CL2;
    Advertising expenditures CL3 > CL2, CL4;
    Trade marketing support CL6 > CL2, CL5, CL1, CL4;
    Size of marketing organization CL3, CL6 > CL5, CL2, CL1, CL4;
    Years of operation CL6 > CL4;
    Sales turnover CL6 > CL5, CL2.

8.5.3 Measures of organizational performance

Three performance measures defined earlier were used to explore whether specific analytically derived patterns of competitive strategy convey performance advantages. Based on ANOVA, the means are presented in Table 8.6. There were no statistically significant differences in the indicators between the groups, that is, within-group variance dominates between-group variance.[7] If we take a look at performance profiles by cluster by comparing the mean performance measures of each cluster with the sample average, the

7. It should be noted that prior empirical evidence linking performance differences with clusters of firms (based either on firm size or on strategy variables) is not very extensive and is conflicting (Thomas and Venkatraman, 1988). The studies have used a wide range of approaches to identify clusters (conceptual and/or method related bases for forming clusters), and hence one cannot compare and aggregate the results across different studies.

Table 8.6 Comparison of organizational performance among six clusters

Cluster	Overall financial performance	Cluster	Efficiency	Cluster	Effectiveness
5	4.8	6	18.5	5	87.7
4	4.3	5	18.3	3	84.2
2	4.1	1	14.2	2	82.9
3	3.8	2	13.4	4	66.3
1	3.8	4	12.4	6	47.5
6	3.4	3	10.2	1	41.9
Average	4.1		14.1		67.8

following observations can be made. With regard to overall financial performance, clusters 3, 1 and 6 fall below the sample average. Cluster 2 appears to have a performance level roughly equivalent to the sample average while clusters 4 and 5 are 'high performers'. Moreover, the efficiency ratio in cluster 6 as well as in cluster 5 is above the sample average, while cluster 3 ranks last. Concerning effectiveness, it can be inferred from Table 8.6 that clusters 5, 3 and 2 have somewhat higher sales growth ratios than is shown by the sample average.

8.6   Conclusions and implications

8.6.1   Evaluating policy effects on the competitiveness of the Finnish food chain

The purpose of the study was to identify the key players, structures and competitiveness in the Finnish food chain, emphasizing policy effects and the competitive advantage of individual firms. In Finland, the important linkage between public policies and agribusiness strategies was accentuated in connection with the major changes due to EU accession and application of the *acquis communautaire*, including the CAP, and entering the SEM immediately from the beginning of 1995. The old national agricultural and food policy did not promote international competitiveness. It rather secured the status quo without providing 'sticks or carrots' to the Finnish food economy to improve profitability and productivity in order to cope with inevitably increasing international competition due to the GATT and EU.

Neither are some of the latest policy developments strengthening the contribution of public policy to formulation of successful business strategies of the food industry and the whole food economy. The national support scheme agreed on in the Accession Treaty is mainly an attempt to secure the established situation, and as such it does not improve efficiency or structural development in Finnish agriculture to any major extent (Kola *et al.*, 1996).

The special transition support granted to food manufacturing companies (e.g. Kettunen and Niemi, 1994; Aaltonen, 1996), mainly in meat and milk processing, should also have had clearer, better targeted restructuring effects in order to gain competitive advantage in the processing stages of the food chain.

Because of the crucial dependence of the Finnish food processing industry on domestic raw material, especially milk and meat, it was quite alarming in the beginning that the support package for the southern agricultural areas which have the highest internal comparative advantage in Finland was not finalized in the Accession Treaty but was left for further consideration. Although the Commission, in Autumn 1996, authorized Finland to grant national aid from 1997 onwards to farmers in southern Finland to alleviate serious remaining difficulties (Article 141 of the Accession Treaty), this region still faces very difficult adjustment problems after the transition support measures cease in 1999. If that leads to a significant decline in production volumes, the threat and incentive to import agricultural and food products in large volumes will increase substantially in Finland. Moreover, poor farm structure and the unfavourable, volatile climate can threaten the domestic supply of raw materials for the food industry in the changing circumstances. These disadvantages cannot be removed by any policy decision, e.g. structural development also takes place only in a longer run.

In addition, the old member countries of the EU-12 now have a greater need to export to the new member states because the GATT Uruguay Round agreement restricts both the value and the volume of subsidized exports to third countries. On the other hand, when the gap between Finnish and EU food prices disappears many EU companies may not regard the small, fragmented, remote and rather overfilled Finnish market as particularly profitable. Moreover, instead of focusing solely on a defensive strategy to keep domestic markets, Finland should also utilize an export-oriented trade strategy in the large, although also quite saturated, EU food markets, and take advantage of its location with respect to market opportunities in Russia and the Baltic states.

8.6.2   Competitive advantages in the food industry: present and future

In the empirical part of the study, a taxonomy of competitive advantages based on cross-sectional data from Finnish food manufacturing firms was developed. The results characterize the situation before joining the EU. Six clusters emerged, providing evidence that a core of groups, each with a distinguishable set of strategies, is discernible. The leading national firms tend to follow either a differentiation strategy in mass markets competing with a wide product range or a distributor-oriented push strategy competing with fewer products. Furthermore, we find a group of medium-sized innovative firms that emphasize a differentiated strategy in regional market segments. There are also a group of smaller firms that have no clear strategic orientation, and a group of very small local firms that are largely production-oriented.

Present strategies in the Finnish food industry are largely home market oriented: distant location and trade barriers, *inter alia*, have contributed to this. The leading firms still have considerable economies of scale in the domestic market. Concerning competitiveness in EU markets, the greatest challenges relate to cost cutting on one hand, and to differentiation on the other. Co-operatives especially are entering a new era of greater uncertainty where the traditions and practices of commodity trading do not work in the long run. It is unlikely that (gradually) opening markets may give new export opportunities for firms producing standard food products because our domestic market is in fact very much smaller than those of EU competitors. In addition, consumer segments are becoming more and more heterogeneous and smaller. In larger firms, product differentiation based on fewer but stronger brands as well as focusing on narrower segments would be a more profitable strategy than the search for economies of scale through wide product ranges. Other options are geographic expansion by acquisition, strategic alliances and co-operation for example through networking. The latter strategy is especially suitable for small firms.

Although price naturally remains important to consumers, quality, safety and ethics of production will also become increasingly important. There is a growing consumer segment that is health and environment conscious and not so price conscious. These trends favour small-scale speciality producers in domestic as well as in foreign markets. In a mature industry such as food, a sustainable competitive position is not only a function of a firm's absolute size. More important are the relative size of the firm in specific market segments and its marketing orientation; product development should rely on consumer needs and preferences rather than on production capabilities. Sustainable competitive advantage is firm-specific rather than industry-specific. In the light of the structure of the European food industry in general, it is unlikely that small-scale and medium-sized food companies will be outcompeted by market leaders. Recently for instance, there have been many entrepreneurial start-ups in prepared consumer foods. While most SMEs may remain domestic, their success may then be largely dependent on the bargaining power of integrated retailers, and the possibilities to choose alternative distribution channels. Direct marketing and selling of products from farms to consumers is one opportunity for SMEs. To conclude, the competitiveness of the Finnish food industry in the international market environment will involve the development of:

- a commitment to innovation and product development
- an organizational culture and capabilities to recognize customer needs and organize cost effective strategies to serve these needs
- focused strategies with a target of achieving dominance in the chosen market area.

Crucial to the success are, not only the strategies of the food industry itself, but also the ability to pass along sufficient incentive to farmers to produce the qualities and quantities of raw materials for processors to meet consumer needs. With respect to the identified six clusters of the food industry, a polarization trend seems to emerge in the Finnish food chain, especially between raw material supply and processing. Firstly, mass-market processors need more efficient, larger farms concentrated in a smaller area in order to create cost-effectiveness through scale economies (as far as it is possible in the Finnish scale of small markets). Secondly, medium-sized innovative firms that emphasize a differentiated strategy in regional market segments require higher quality and time-responsiveness of farms. One key success factor allowing the development of differentiated and/or speciality products is the co-operation between retailers, food manufacturers and farmers. Contract production, quite rare up to now in Finland, is bound to increase in both mass market and specialty products. It means that advantages of vertical integration have to be better utilized in terms of the comprehensive systems of logistics, including novel attempts at creating information chains describing the origins, production practices, etc. of raw material and food products from farmers to consumers. Horizontal integration of farms through co-operation is also essential to acquire cost-effectiveness, which may give farmers some bargaining power in the food chain.

8.6.3 Policy implications

Below we briefly outline a general proposal on what kind of existing public policies could be applied to promote successful development of the firm clusters we identified in the study. These policy measures should be able to deal with the evident need to modify strategies and to alleviate the problems that the Finnish food industry faces in a changing environment and increased foreign competition.

- **Cluster 1** (large firms of distributor-oriented, low-price strategy operating regionally and nationally) and **cluster 6** (very large, old firms of distributor-orientation and product development, a push strategy group).
  Policy choice: General economic policy, including financial and monetary policies, that enhance favourable economic development (growth but low factor costs) and investment possibilities is important for these larger firms. Trade policies of trade creation could also be beneficial for these firms as they are the potential exporters based on their current size. In addition, R&D policy can enhance the product development strategies prevailing in cluster 6.
- **Cluster 2** (SMEs utilizing differentiation and cost-based advantages without clear strategic orientation, a so-called 'stuck in the middle' group), **cluster 3** (large firms of marketing differentiation lacking cost-

282 COMPETITIVENESS IN THE FOOD INDUSTRY

based advantages) and **cluster** 5 (small firms of production and cost-oriented strategies lacking marketing competence).
Policy choice: in terms of efficiency of policy and budget spending, no specific policy can or should be devoted to 'rescuing' these groups as they are in any case gradually quitting due to lack of earlier development and future management visions (clusters 3 and 5), and due to unsustainable choices made with respect to changing environment (cluster 3: broad product range but no cost advantage).

- **Cluster 4** (innovative SMEs lacking cost-based advantages.
  Policy choice: R&D policy to help innovation. Regional policies, including also the structural funds of the EU (e.g. objective 5b) to enhance and develop further the possible specific strengths in raw material supply and firm co-operation in the region of most potential markets. Environmental policy to focus on high quality of the raw material to make it possible to differentiate products according to that strength. Financial policy, e.g. low interest rates and investment aids, to maintain possibilities to invest in new products and their development, but also to improve cost-efficiency of these firms through expansion. Strict competition policy should also help this group, considering the stronger market power of clusters 1 and 6 of larger firms.

As a whole, agricultural policy is generally important to the food industry, but it is crucial for Finnish milk and meat processing firms as it strongly affects domestic raw material supply to these firms. Hence, as we identified cost inefficiency and poor structure as problems for primary production, structural policy including investment aids and general economic policy, with favourable taxation decisions and interest rates, for example, could help. Moreover, competition policy could promote, or at least allow, tighter integration in the entire food chain in order to improve its efficiency. On the other hand, if the slow adjustment ability of agriculture to changing consumer preferences (ecological and ethical concerns) appears as a problem, agri-environmental policy to redirect production methods, R&D funding for finding and adopting new products, markets, and technology, and 'information' policy to educate both producers and consumers could be employed. Creation of the so-called information chains from fields to tables is one example of the latter alternative, supported by government, farmers and food firms, and strongly preferred by consumers. The aim is to reduce the problems caused by incomplete and asymmetric information.

The aforementioned measures can represent decisive redirections in policy. Before 1995, in spite of overproduction and export problems prior to EU membership, there were no policy changes that would have facilitated the reorientation of both farmers and food manufacturers to the challenges of international competition, for example market-led strategies emphasizing product differentiation, or the development of ecological production.

Instead of too heavy reliance on subsidies, agricultural policy should have utilized more comprehensive education and extension programmes to introduce entrepreneurial/management thinking among farm families to improve their basic income formation, perhaps through adoption of new technology to increase value added, or to find new income sources through the diversification of farm activities.

In terms of changes in consumer needs and preferences, one possibility to improve the competitiveness of the Finnish agriculture and food industry is the new comparative advantage, which has its foundations in the operations and (natural) environment of production that are perceived to be acceptable by public opinion (Kola, 1995). Food safety and security (also economics of safety and security to an increasing extent), ethics of production, and quality of many and diverse dimensions are decisive. Sustainability in general in agricultural and food production could be the issue, in which public policy programmes can – and clearly must– respond to consumer preferences, and thereby improve competitiveness in the entire food economy. In a small country like Finland a differentiation strategy according to the new comparative advantage could also apply at the national policy level (Brink and Kola, 1995). It can be aimed at distinguishing the whole small country's industry/product segment through eg. national branding from that of its (large) competitors. It can complement, but possibly also re-direct the strategies of the food industry. The working group on agricultural policy, led by the minister of agriculture and forestry, includes the creation of a quality system for the entire food chain in its key proposals to improve competitiveness in Finnish agriculture and the entire food economy (MMM, 1996). Here, public policies and business strategies actually can converge, and public policy can fulfill its objective to increase overall welfare in society.

References

Aaltonen, S. (1996) Adjustment of the Finnish food industry, in *First Experiences of Finland in the CAP* (ed. L. Kettunen), Research Publications 81, Agricultural Economics Research Institute, Helsinki, pp. 43–56.

Azzam, A.M. and Schroeter, J.R. (1995) The tradeoff between oligopsony power and cost efficiency in horizontal consolidation: an example of beef packing. *American Journal of Agricultural Economics*, 77(4), 825–36.

Bain, J.S. (1968) *Industrial Organisation*. Wiley, New York.

Barney, J. (1991) Firm resources and sustained competitive advantage. *Journal of Management*, 17(2), 99–120.

Brink, L. and Kola, J. (1995) Small countries with large neighbours: choosing agri-food policies to improve competitive performance. Discussion group report, XXII International Conference of Agricultural Economists, Zimbabwe. *IAAE Members Bulletin*, 13, May.

Brownlie, D. and Moutinho, L. (1990/91) Competitive analysis in strategic marketing: specious or substantive? *Irish Marketing Review*, 5(1), 39–49.

Conner, K.R. (1991) A historical comparison of resource-based theory and five schools of thought within industrial organization economics: do we have a new theory of the firm? *Journal of Management*, 17(1), 121–54.

284 COMPETITIVENESS IN THE FOOD INDUSTRY

Cravens, D.W. (1987) *Strategic Marketing*, 2nd edn. Richard D. Irwin, Homewood, IL.
Day, G.S. and Wensley, R. (1988) Assessing advantage: a framework for diagnosing competitive superiority. *Journal of Marketing*, 52(April), 1–20.
Day, G.S. (1990) *Market Driven Srategy: Processes for Creating Value*. Free Press, Collier Macmillan, London.
Dess, G.G. and Davis, P.S. (1984) Porter's (1980) generic strategies as determinants of strategic group membership and organizational performance. *Academy of Management Journal*, 27(3), 367–488.
Dess, G.G. and Robinson, R.B. (1984) Measuring organizational performance in the absence of objective measures: the case of privately-held firm and conglomerate business unit. *Strategic Management Journal*, 5, 265–73.
Everitt, B.S. (1980) *Cluster Analysis*. Heinemann Educational, London.
Eaton, J. and Grossman, G.M. (1986) Optimal trade and industrial policy under oligopoly. *Quarterly Journal of Economics*, 101, 383–496.
Finnish Food and Drink Industries (1995) *Facts about the Finnish Food Industry*. Finnish Food and Drink Industries, Helsinki.
Gassman, H. (1996) Globalisation and industrial competitiveness. *OECD Observer*, 197, 38–42.
Hardy, K.G. and Magrath, A.J. (1988) *Marketing Channel Management, Strategic Planning and Tactics*. Scott, Foresman, Glenview, IL.
Henson, S., Loader, R. and Traill, B. (1995) Contemporary food policy issues and the food supply chain. *European Review of Agricultural Economics*, 22(3), 271–81.
Hernesniemi, H., Lammi, M. and Ylä-Anttila, P. (1995) *Kansallinen kilpailukyky ja teollinen tulevaisuus*. Elinkeinoelämän Tutkimuslaitos ETLA/Suomen itsenäisyyden juhlarahasto SITRA, Taloustieto Oy, Helsinki.
Hill, C.W. (1988) Differentiation versus low cost or differentiation and low cost: a contingency framework. *Academy of Management Review*, 13(3), 401–12.
Hunt, S.D. and Morgan, R.M. (1995) The comparative advantage theory of competition. *Journal of Marketing*, 59(April), 1–15.
Hyvönen, S. (1993) *The Development of Competitive Advantage: An Identification of Competitive Strategy Patterns in Food Manufacturing Firms*. Publication B-131, Helsinki School of Economics and Business Administration, Helsinki.
Hyvönen, S. and Kola, J. (1995) The Finnish food industry facing European integration: strategies and policies. *European Review of Agricultural Economics*, 22(3), 296–309.
Jones, W.D. (1995) Competition Policy and the Agro-food Sector. Paper presented at the 44th EAAE Seminar, October 1995, Thessaloniki, Greece.
Kettunen, L. (ed.) (1996) *First Experiences of Finland in the CAP*. Research Publication 81, Agricultural Economics Research Institute, Helsinki, pp. 143–157.
Kettunen, L. and Niemi, J. (1994) *The EU Settlement of Finnish Agriculture and National Support*. Research Publication 75a, Agricultural Economics Research Institute, Helsinki.
Kim, J. and Mueller, C.W. (1978) *Factor Analysis: Statistical Methods and Practical Issues*. Sage University Press, Beverly Hills, CA.
Kola, J. (1993) Principles of agricultural policy in Finland in relation to the CAP of the EC, in *Finnish Agriculture and European Integration* (ed. L. Kettunen), Research Publication 71, Agricultural Economics Research Institute, Helsinki. pp. 21–36.
Kola, J. (1995) Agribusiness caught between efficiency and environment: the role of a small country in the game of large countries, in *Managing in the Global Economy* (ed. W.D. Gorman), IAMA, Texas A&M University, pp. 336–46.
Kola, J., Hyvönen, S. and Vironen, T. (1996) Agriculture and food industry: convergence between public policies and business strategies, in *Agro-Food Small and Medium Enterprises in a Large Integrated Economy* (eds. K. Mattas, E. Papanagiotou and K. Galapanopoulos) Proceedings of the 44th EAAE seminar, Wissenschaftsverlag Vauk, Kiel, pp. 16–28.
Laurila, I.P. (1996) Consequences of the EU membership on food prices and food consumption in Finland. *Working Paper* 1/96, Agricultural Research Institute, Helsinki.
Mahoney, J.T. and Pandian, J.R. (1992) The resource-based view within the conversation of strategic management. *Strategic Management Journal*, 13, 363–80.
Marttila, J. (1996) *The Effect of Oligopolistic Competition on Economic Welfare in the Finnish Food Manufacturing*. Research Publication 80, Agricultural Economics Research Institute, Helsinki.

McNally, M. (1995) *International Wheat Subsidies: Who Really Profits?* Garland Publishing, New York.

Miller, D. and Friesen, P.H. (1984) Archetypes of strategy formulation. *Management Science*, 24, 921–33.

MMM (1996) The agricultural policy working group, intermediate report. *MMM 1996:14*, Ministry of Agriculture and Forestry, Helsinki.

Montgomery, C.A. and Porter, M.E. (1991) *Strategy: Seeking and Securing Competitive Advantage*. Harvard Business School Publishing, Boston.

Murray, A.I. (1988) A contingency view of Porter's generic strategies. *Academy of Management Review*, 13(3), 390–400.

Nelson P.R. and Winter S.G. (1982) *An Evolutionary Theory of Economic Change*. Belknap Press, Cambridge, MA.

Nunnally J.C. (1978) *Psychometric Theory*, 2nd edn. McGraw-Hill, New York.

Olver, J.M. and Harris, P.W. (1989) Push and pull: a one–two punch for packaged products. *Sloan Management Review*, Fall, 53–61.

Penrose, E. (1959) *The Theory of the Growth of the Firm*. Wiley, New York.

Porter, M. (1980) *Competitive Strategy*. Free Press, New York.

Porter, M. (1985) *Competitive Advantage. Creating and Sustaining Superior Performance*. Macmillan, New York.

Porter, M. (1990) *The Competitive Advantage of Nations*. Macmillan, New York.

Porter, M. (1991) Towards a dynamic theory of strategy. *Strategic Management Journal*, 12, 95–117.

Prahalad, C.K. and Hamel, G. (1990) The core competence of the corporation. *Harvard Business Review*, May–June, 79–91.

Punj, G. and Stewart, D.W. (1983) Cluster analysis in marketing research: review and suggestion for application. *Journal of Marketing Research*, 20, May, 134–48.

Reibstein, D. (1995) Do marketing expenditures to gain distribution cost the customer? *European Management Journal*, 13(1), March, 31–7.

Robinson, R.B. and Pearce, J.P. (1988) Planned patterns of strategic behavior and their relationship to business-unit performance. *Strategic Management Journal*, 9, 43–60.

Romppanen, A. and Leppänen, S. (eds.) (1995) *The Global Economy and Finland*. VATT Publication 21, Government Institute for Economic Research, Helsinki.

Rumelt, R.P. (1987) *Strategy, Structure and Economic Performance*, 2nd edn. Division of Research, Graduate School of Business Administration, Harvard University, Cambridge, MA.

Scherer, F.M. and Ross, D. (1990) *Industrial Market Structure and Economic Performance*. Houghton Mifflin, Boston.

Selznick, P. (1957) *Leadership in Administration*. Harper and Row, New York.

Shaw, S.A., Dawson, J.A. and Blair, L.M. (1992) The sourcing of retailer brand food products by a UK retailer. *Journal of Marketing Management* 8(2), 127–46.

Stern, L.W. and El-Ansary, I.A. (1992) *Marketing Channels*. Prentice Hall, Englewood Cliffs, NJ

Thomas, H. and Venkatraman, N. (1988) Research on strategic groups: progress and prognosis. *Journal of Management Studies*, 13(3), 537–55.

Tirole, J. (1988) *The Theory of Industrial Organization.* MIT Press, Cambridge, MA.

Venkatraman, N. and Ramanujam, V. (1986) Measurement of business performance in strategy research: a comparison of approaches. *Academy of Management Review*, 11(4), 801–14.

Viaene, J. and Gellynck, X. (1995) Structure, conduct and performance of the European food sector. *European Review of Agricultural Economics*, 22(3), 282–95.

Volk, R., Laaksonen, K. and Mikkola, H. (1996) *Säätelystä kilpailuun: elintarvikeklusterin kilpailukyky*. Pellervon taloudellinen tutkimuslaitos, raportteja ja artikkeleita No 140. Espoo.

Wernerfelt, B. (1984) A resource-based view of the firm. *Strategic Management Journal*, 5, 171–80.

9 Are Porter diamonds forever?
MAGNUS LAGNEVIK AND JUKKA KOLA

9.1 Introduction

The Porter (1990) model has been widely used and much discussed since it was introduced. In Chapter 1 we give references to some of the most interesting contributions to the academic debate and in the appendix to that chapter we list important European studies of competitiveness in the agri- and food industries which have used the Porter diamond model.

In this concluding chapter we summarize our own experience from using the model in six European countries. Each study used Porter (1990) as the point of departure. However, the model has been applied in slightly different ways in the various studies. In this way, this book represents an attempt to develop the Porter model further and to learn from applications in new areas. In this final chapter we describe our experience of the practical empirical work with the model and discuss the implications for the theoretical model. We discuss the benefits and shortcomings with the model and the extent to which our results support the Porter model or indicate a need for alternatives.

In section 9.2 we discuss the relationship between competition and competitiveness. In section 9.3 the focus is on the home base concept and its meaning. Section 9.4 deals with the role of retailing in the food chain, and section 9.5 is concerned with the role of the EU Common Agricultural Policy (CAP) in the competitiveness of the food chain in various countries. Section 9.6 investigates the dynamics of change and how the dynamics can be increased, and in section 9.7 we describe how we have applied the Porter model in new ways and suggest areas for future research.

One important conclusion that we have reached is that the Porter model has worked well as a research tool. Although we have run into various difficulties in our attempts to use the model, and have debated intensely among ourselves how well the model works, where it is applicable and where not, nevertheless we – a dozen researchers with different theoretical backgrounds from six countries – have managed to create a constructive academic debate. Furthermore the project has produced interesting results. We have changed the basic working model for the analysis of the performance of an industry from the often used static one – structure–conduct–performance (SCP) – to a dynamic one. New issues have been raised as a result of our studies. In this sense we conclude that the Porter (1990) model has worked as a common theoretical framework. It has enabled interdisciplinary work and it has generated new research ideas.

9.2 Competition and competitiveness

The European scene in the agri-business and food industry has changed dramatically in the 1990s. Competitive pressure has increased through rationalization within Europe and through increasing competition created by the GATT agreement. The discussion of the funding of the EU agricultural budget and the ongoing addition of new member states has created an atmosphere where it is now widely accepted that regulations and political control will have to give way to markets, competitiveness and increasing efficiency.

In this new situation, many players in the market have reached the conclusion that bulk production will lose profitability if price supports are reduced. Very few think that in the future the EU can devote large sums to price support. Some argue that in the EU in future the focus of attention will be the cities, not the rural areas (Delebarre, 1992). Whether you are in farming, food processing, wholesaling or retailing, increased price pressure and continuous competition seem to form a realistic future. Therefore, in building competitiveness, the creation of value added is seen as a very important part of corporate and industry strategies.

Value added is created through use of high quality raw material, and through use of technology in product development and processing of food. Brand names and quality certificates are also important in communicating product quality to consumers.

In the value added chain for food products there are many important linkages between the different chain stages. for the firm that wants to build competitiveness through added value, it can be vital to co-operate with other companies. In the Swedish ecological case the importance of well organized supply chains from producers to retailers was demonstrated. For the retailers the KRAV label, indicating supply from a defined group of farms with a certified production method, was necessary in order to give credibility to the ecological market strategy. For the producers, long term relationships with retailing were necessary in order to get the sustainable volumes that made it possible to invest in distribution and certification systems. In contrast, we see that the failure of the UK mushroom and apple producers and processors to develop good relationships with research institutes, so that advanced British scientific knowledge could be incorporated into British products, has created a loss of competitiveness for these sectors. The case for co-operation is also well illustrated by the successful Italian agri-food districts and it is proposed in Chapter 5 that co-operation could help Belgian pigmeat companies (see also Fanfani and Lagnevik, 1995, p. 22). In Ireland the dairy industry successfully co-operates in export ventures. Thus there are visible signs of the use of simultaneous competition and co-operation as both corporate strategies and industry strategies throughout the EU.

This leads us to the conclusion that we disagree with what seems to be both a basic assumption and a main result of the Porter model – that competition alone is the driving force that explains competitiveness. In our studies, the ability to co-operate has been very important in the building of competitiveness, although that co-operation takes place in an increasingly competitive environment. To leave co-operation out of the picture could mislead competition authorities into banning long term contracting and other efforts which are essential ingredients in the building of competitiveness through value added products in Europe.

We have concluded, however, that the Porter model can be very well used without making the assumption that the system is necessarily driven by competition. In this respect we are in agreement with Enright (1990, 1994).

9.3 The home base

What is the home base of an industry? The definition of the home base concept has been problematic throughout our studies. This can be illustrated by way of a few examples.

Initially we discovered that the clusters we investigated did not correspond well with national borders. To mention obvious cases, the conditions for agricultural production are very different in northern Finland and northern Sweden compared to the southern parts of these two countries. In Italy the clusters were concentrated in regions, and in Ireland there are some very important linkages with retailing in the UK. In addition, we must recognize that in many important respects the EU is one single market.

For at least three of the countries in the study – Finland, Ireland and Sweden – the home market is very small. In some cases of export success, such as Ireland, it seems that the home market demand is not very important, not very sophisticated and not a driving force in the building of competitiveness.

In at least two cases, Italy and Sweden, we find that the home market conditions have been very important for the development of pasta products and ecological food products respectively. What is more troublesome is the successful Irish case where success is created from small and unimportant home markets, though here we can see the importance of dedicated government in the development of an industry of great importance to the nation.

From a conceptual point of view, it is useful to distinguish between the home base environment for research and development, skilled labour, specialized infrastructure and local networks on one side and the market- and customer-oriented networks on the other side. We need to distinguish between home market and home base. These concepts are not identical. Home market refers to the existence of sophisticated buyers who can anticipate future buyer needs in other countries. Home base concerns all the components of the Porter diamond, and the way they interact with each other. Porter (1990, p. 606) argues that a firm can only have one true home base. If it tries to have several, it will sacrifice the dynamics that arise from true integration in a national diamond. Globalization in a corporation should therefore not be used to create several home bases, but should aim at tapping selectively into sources of advantage in other diamonds. According to this line of argument we see the need for two kinds of networks or clusters for the corporation. One deals with factor creation and development issues and is often regional–national. The other one is international and is of importance for the market intelligence function and for the acquisition of competences that are not available in the home base.

There are three principal features of regional clusters that influence firm strategy (Enright, 1994). The first is that the resources and capabilities vital for the success of firms can often be found inside a region rather than inside a single firm. The second is that regional clusters often involve activities that are shared between firms within a cluster. The third realization is that firm strategic choices can be influenced by strategic interdependencies, rapid information flows, and the unique mix of competition and co-operation often found in regional clusters.

The growth and persistence of regional clusters results from the pressures, incentives and capabilities to innovate provided by the local environment (Enright, 1990). Often informal and oral information sources provide key information about market needs and technological possibilities that lead to innovation (Utterback, 1974). It has been suggested that specialist or industry specific information is subject to steep distance decay (Goddard, 1978) though modern global communications must diminish this effect. Regional clusters often provide the stimulus for public and private investments. Local industry associations provide commercial research of foreign markets. Local governments often make contributions to industry-specific infrastructure. Local universities often provide industry-specific research and specialized training. Geographic concentration provides firms within a cluster with short feedback loops for ideas and innovation.

The second kind of network or clusters for the innovative firm adds to the geographically limited network centred around production, technology, unique resources and specific infrastructure. This network, which is often global, includes the key customers in foreign markets (Kristensen, 1992). The contribution of such partners, e.g. distributors, retail chains and industrial customers to the product development process can be vital. According to Langhoff (1993), the importance of a partner's supplementary competencies for the food company's internationalization process increases, as cultural heterogeneity in the market increases (Grunert et al., 1996)

Thus regional and global perspectives are not contradictory, but complementary. In the agri-food value added chain there are many interdependencies between the various parts of the chain. Product quality as well as product uniqueness can often be dependent on co-operation between the different parts of the chain, wherever they are located. It is therefore no contradiction to say that the regional cluster is essential and at the same time to say that international clusters are vital in the upgrading of competitiveness.

Interactions within a cluster can also be within parts of nations. Even if Porter started from national statistics when identifying successful clusters, this does not mean that all parts of a cluster are national and only national. We now return to our original question. How do we define a cluster? We turn to Beccatini (1989) for help and investigate his definition of an industrial district as a localized thickening of interindustrial relations with reasonable stability over time. In order for a district to remain stable over time, there must exist 'a complex and inextricable network of external economies and diseconomies'. These factors are to be understood as external to the enterprise but internal to the cluster.

This provides us with a functional definition of the cluster. We turn away from the question of how it corresponds to national borders and focus our attention on what happens in a cluster and what functions are performed. Here we can also find a contribution from Porter when he discusses interchange within clusters (1990, p. 153):

> Mechanisms that facilitate interchange within clusters are conditions that help information to flow more easily, or to unblock information as well as facilitate co-ordination[1] by creating trust and mitigating perceived differences in economic interest between vertically or horizontally linked firms. Some examples are the following:
> *Facilitators of information flow*
> • Personal relationships due to schooling, military service
> • Ties through the scientific community or professional associations
> • Trade associations encompassing clusters
> • Norms of behaviour such as belief in continuity and long term relationships
> *Sources of goal congruence or compatibility within clusters*
> • Family or quasi-family ties between firms
> • Common ownership within an industrial group
> • Ownership of partial equity stakes
> • Interlocking directors
> • National patriotism.

If we want to operationalize the 'localized thickening of interindustrial relations with reasonable stability over time', we can turn to Porter and get examples of phenomena that can be measured in order to find clusters. By combining the Porter approach with results from the ID school of thought and also with Grunert *et al.* (1996), we can make a better portrait of the home base and show what needs to be investigated in order to identify it.

9.4   Retailing and competitive dynamics

Retailers have become dissatisfied with their role in the value added chain. Their traditional role as distributor of goods from producers to consumers

---

1. This is the closest Porter ever gets to the concept of co-operation, a word he never once uses in his book.

has been proven to be too limited and to provide too little earnings. As a result retailers have started their own product development, developed their own brands and developed a number of services connected to food products. The distinctions between retail stores, catering services and restaurants are hard to find nowadays.

One effect is that food processing companies are forced to review their cost structure. If an order to deliver products for private label is accepted, the producing company is able to cut out marketing costs. It must deliver according to specification and have a cost competitive production structure. The alternative strategy is to build brands that are so well recognized by customers that retailers need to include them in their assortment.

In this situation new ways to compete have developed. Alliances in product development between retailers and food processing companies are quite common (Traill and Grunert, 1997). The aim of these alliances – medium or long term – is to develop unique products that will outcompete the products of other retail chains. For the retailer it is important to find partners with product development capacity and cost effectiveness. For the food processing company it is vital to choose the right outlet, so that volumes and margins are sufficient to create good results.

Another version of this kind of co-operation is **efficient consumer response** (ECR) (Thomas *et al.*, 1995, Fiorito *et al.*, 1995), where the ambition is to create an integrated information handling and distribution system from producer to final consumer. The three pillars of ECR are to provide consumer value, to remove costs that do not add value and to maximize value and minimize inefficiency throughout the supply chain. To implement ECR, distributors and suppliers reduce inventory, eliminate paper transactions and streamline product flow.

Order cycles are shortened through the use of technology, primarily electronic data interchange and barcodes to automatically identify products. Order cycles are further reduced through 'strategic partnerships', where retailers and manufacturers work together as a team to establish ways of achieving performance goals that exceed existing industry practices. Not only is the order cycle addressed, but so also are a wide variety of business processes involving new product introductions, item assortments and promotions.

Participation in these co-operative efforts between retailers and food processing companies requires a critical mass of operations and competence. The Irish research team noticed in the their case study that large food processing companies report positive experience from interaction with retailers, while small companies see retailers mainly as difficult customers who reduce processors' margins.

In the Swedish case we notice the importance of marketing strategy of the three major retail chains for ecological producers. The combination of an efficient distribution organization, certification and private labels were key elements in the creation of increased market shares for ecological products.

The fact that Swedish retailing is quite concentrated helped the new products to reach the market quickly.

In the Italian case we find the opposite situation. The small scale structure of Italian retailing was a restriction on the creation of competitive dynamics in the home market. Irish dairy manufacturers gained some advantage from dealing with Irish retailers which was useful for entry to the UK market, but British horticultural industries gained little from direct contact with Europe's most sophisticated retailers. It must be remembered that an outward looking retail system is more likely to indulge in global sourcing of products

Thus we have a new kind of competition where the consumer meets several new product concepts. New food products with high value added find their way to consumers with increasing frequency. But these products are often developed in co-operation between several actors along the value added food chain. In the creation of new food products the changing role of the retailers is an important force creating competitive dynamics. These dynamics also call for reviews of production, marketing, distribution and development strategies in the whole value added chain.

9.5 Competitiveness and the CAP

In the Porter model, policy is conventionally represented through 'government', an outside influence on the four determinants of the diamond. However, we find in our case studies, and want to emphasize for future consideration, that policy/government does not influence the Porter diamond from outside only. In particular, policy can also be affected, and policy makers and industry representatives can co-operate in order to improve competitiveness of a sector to benefit the entire economy.

In the 1990s, competitive pressures have grown in food markets due to trade and agricultural policy reforms at both international (GATT Uruguay Round 1986–1994) and EU levels (the creation of the single European market and CAP reform in 1992). Consequently, increased attention has been paid to the linkage between public policy and competitiveness of the agri-food industries (Brink and Kola, 1995; Henson *et al.*, 1995; Jones, 1995; Kola *et al.*, 1996; see also the Finnish and Irish case studies in this book).

The agricultural policy of the EU (the CAP) strongly affects the functioning of the European food industry.[2] The 1992 reform of the CAP was still insufficient to fundamentally redress key problems: the 1995 arctic–alpine

---

2. However, there are many other policy regimes that affect different industries; some affect all industries horizontally (e.g. technology, research, trade policy), some certain industries almost individually (e.g. the CAP and the agri-food industry). For a better description, see the Finnish case study in Chapter 8.

enlargement of the EU brought new problems; and in the face of the extensive effects of the Eastern enlargement, further and more radical reforms are necessarily needed (Kola, 1996).

Although the CAP is the main factor in the European agri-food sector, its effects appear quite different for the countries we studied. The Irish case shows how the CAP has had a strong positive effect on the success of the dairy industry in a country with a small home market. The CAP regime has clearly discriminated in favour of the Irish dairy sector, both at the farm and processing level, though some of this favourable treatment is apparently a result of successful organization and lobbying of the Irish industry to influence the CAP. This finding is in line with Kamann and Strijker (1995, p. 107), who argue in their study of the Dutch dairy industry that the CAP is not something autonomous and given, but is politically determined and therefore subject to lobbying. This is, of course, only a repetition of common knowledge in the political economy literature. However, there is still a need for better understanding of why and how certain interest groups and/or countries, like Ireland here, are more successful than others in influencing political decision-makers. Naturally, a strong dependence of a country on the agri-food industry, and especially its export earnings, increases the incentive to lobby, at all levels, for the national interest, not only those of firms or industries.

In the other country cases, the effects of the CAP are clearly smaller, if they exist at all. The Finnish case in particular illustrates the unsuitability of the CAP regime, in spite of its complexity, for the very different agricultural production conditions found in Finland (e.g. unfavourable climate and farm structure). The same policy that has provided big benefits for some countries, here Ireland in particular, poses major difficulties for others such as Finland. The fundamental policy problem here is that the 'Common Agricultural Policy' is not common enough to meet the needs of increasing number of different member states. The eastern enlargement will be the ultimate test for the CAP, and a fundamental reform is inevitable.

The Swedish case, in turn, illustrates the inability of the CAP to respond sufficiently quickly to the changing preferences and new needs among consumers, or societies at large. Ecological consumer products require ecological/organic raw material from primary agricultural production, which is not sufficiently well covered, controlled or encouraged in the current CAP. To some extent, the agri-environmental programmes of the CAP can contribute to the improvement of competitiveness in the food industry by developing, possibly even differentiating, production techniques in agriculture, and thereby increasing value added in the entire food chain.

In the Italian case the effects of the CAP are somewhat ambiguous, although government policy concerning the retail structure has had a negative effect for the development of competitiveness. In the UK case the actions of national government have clearly not promoted competitiveness in the horticulture sector. The Belgian case shows that the necessity for the CAP to follow the GATT obligations of reduced export subsidies will indirectly result in the deterioration of the competitive position of a relatively poorly developed Belgian meat industry. Because of GATT obligations, EU producers will be forced to redirect their meat products from exports to third countries on to the internal market, i.e. to compete with the traditional markets of Belgium in neighbouring EU states.

Continued deregulation and declining support are inevitable for the CAP as a whole, including the dairy regime which remained relatively untouched in the 1992 CAP reform. Clearly, the CAP has not encouraged the Irish industry to develop strategies, in terms of product portfolio and industry structure, for instance, to succeed in the future of stronger competition from more efficient producers in the EU (e.g. the Netherlands and Denmark) and in third countries (New Zealand).

The ability of current agricultural policy to promote the long term sustainability or competitiveness of the European food industry appears quite limited in our studies. More specifically, in the Irish and Finnish cases we can conclude that agricultural policy aimed at providing security and stability to the sector has reduced, if not totally eliminated, the need to develop new strategies in the agri-food industry in order to cope with the changing environment and increasing international competition. In spite of its extensive effects on the food industry and consumers, the CAP has been seen too much as a purely agricultural concern (e.g. Swinbank and Harris, 1991). This has further hindered the development of appropriate actions and strategies to improve competitiveness. It is of course evident that as the share of value added increases throughout the food industry, the role of agricultural policy will become less fundamental than in the past when food was basically a series of commodity products.

9.6   Change of industry dynamics

In several of the cases that we have studied, the dynamics of the industry have changed. In the Belgian case, the structure of the meat processing industry needs to improve to respond to changing demand and policy reform and remain cost effective. In the UK, changes in retailing and in the nature of international research mean that more must be done for the mushrooms, apple and strawberry industries to regain competitiveness. In Ireland the successful dairy industry has ideas about the strategic changes that may be necessary in order to adapt to a future where the CAP is reformed and has only minor importance for the agri- and food industry. In the Finnish and Swedish cases the question is how to adapt to a situation as new members of the EU.

In a more integrated Europe with fewer barriers we can expect industrial agglomerations to occur in the locations where the infrastructure, the industrial climate and competitive dynamics are well developed, regardless of national boundaries (Marshall, 1966, Krugman, 1991). We also know that national measures to limit the competitiveness of companies of a different nationality are not acceptable in a single market.

What remains for the regions and the nation states is to develop the infrastructure, the education, the research and development, the climate for innovations and the incentives to create and improve agri- and food companies.

We regard it as vital to understand competitive dynamics in a specific setting in time and space. Industrial policy should therefore be less general and more adapted to specific industries. It should also be clear that conditions for efficient industries vary between region and that we can expect industry specific industrial agglomeration to increase.

9.7 Porter: methodological development

In this book we have developed Porter's concepts further. We have a different view to Porter when it comes to the role of competition and co-operation in the competitive dynamics, but we find his model useful in spite of these differences. We have discussed and redefined the concept of home base, adapting it to conditions more valid for the European agri- and food industries. Furthermore we have analysed sources of competitiveness.

The Finnish case study demonstrates concepts and methodologies for measurement of such sources. An empirical quantitative research method is presented and used. In the Belgian case a development of the Boston Consulting Group model for industry competitiveness analysis is presented and used. The Irish case illustrates a complete Porter analysis where quantitative measures of competitive advantage – revealed competitive advantage – supplements a qualitative analysis of industrial dynamics in a new way for this kind of study. The Swedish case uses the Porter approach to analyse an emerging market to see if an industry can reach international competitiveness in the future. The UK case applies a Porter analysis to an uncompetitive industry. The Finnish case analyses the role of different kinds of policy for industry strategies adapted to the situation in the agri- and food industry. This study provides a framework and an analysis of policy that goes beyond Porter.

Our overall conclusion is that Porter has proved a useful for the analysis of international competitiveness. In particular it provides a checklist of items that should be considered and a guide to the ways in which they may successfully interact to create a competitive dynamic. The method highlights the value of qualitative analysis applied through the case study approach. It is not the whole story, however, and beyond this book, we would like to encourage colleagues to conduct European and international comparative studies of competitive industrial clusters. Such studies would provide us with valuable information about characteristics specific to the country, region and industry of the processes in which competitiveness is upgraded.

References

Becattini, G. (ed.) (1989) *Modelli locali di sviluppo*. Bologna, Il Mulino, p. 231.
Brink, L. and Kola, J. (1995) Small countries with large neighbours: choosing agri-food policies to improve competitive performance. Discussion group report, XXII International Conference of Agricultural Economists, Zimbabwe. *IAAE Members Bulletin*, 13, May.
Delebarre, M: (1992) *Information Technology, an Opportunity for the Cities. Cities and New Technology*. OECD, Paris.
Elg, U. and Johansson, U. (1995) *Prevailing Distribution Networks: barriers or mediators of european integration*. Institute for Economic Research Working Paper Series 1995:18, School of Economics and Management, Lund University.
Enright, M: (1990) Geographic concentration and industrial organization. PhD dissertation, Harvard University.
Enright, M: (1994) Organization and co-ordination of geographically concentrated industries, in *Co-ordination and Information. Historical Perspectives on the Organization of Enterprise* (eds D.M.G. Raff and N.R. Lamoreaux). Chicago University Press, Chicago.
Fanfani, R. and Lagnevik, L. (1995) Industrial districts and Porter diamonds. Paper prepared within the EU Concerted Action Structural Change in the European Food Industry, presented at the Strategic Management Society 15th Annual Conference, Mexico City, 15–18 October.
Fiorito, S., May, E. and Straughn, K (1995) Quick response in retailing: components and implementation. *International Journal of Retail and Distribution Management*, 23(5), 12–21.
Goddard, J: (1978) The location of non-manufacturing activities within manufacturing industries, in *Contemporary Industrialisation* (ed. F. Hamilton), Longman, London.
Green Paper on Innovation (1995) European Commission, Brussels.
Grunert, K. *et al.* (1996) *Market Orientation in Food and Agriculture*. Kluwer, Boston.
Henson, S., Loader, R. and Traill, B. (1995) Contemporary food policy issues and the food supply chain. *European Review of Agricultural Economics*, 22(3): 271–81.
Jones, W.D. (1995) Competition policy and the agro-food sector. Paper presented at the 44th EAAE Seminar, Thessaloniki, Greece, October.
Kamann, D. and Strijker, D.(1995) The Dutch dairy sector in a European perspective, in *The Dutch diamond: The Usefulness of Porter in Analysing Small Countries* (eds P.R. Beije and H.O. Nuys), Garant, Leuven-Apendoorn.
Kogut, B. (1993) *Country Competitiveness*. Oxford University Press, Oxford.
Kola, J. (1996) From the CAP to a RAP, in *First Experiences of Finland in the CAP* (ed. L. Kettunen), Research Publication 81, Agricultural Economics Research Institute, Helsinki, pp. 143–57.
Kola, J., Hyvönen, S. and Vironen, T. (1996) Agriculture and food industry: convergence between public policies and business strategies, in *Agro-Food Small and Medium Enterprises in a Large Integrated Economy* (eds K. Mattas, E. Papanagiotou and K. Galapanopoulos). Proceedings of the 44th EAAE seminar, Wissenschaftsverlag Vauk, Kiel, pp. 16–28.
Kristiansen, P.S. (1992) Product development strategy in the Danish agricultural complex: global interaction with clusters of marketing excellence. *International Journal of Food and Agribusiness Marketing*, 4(3) 107–18.
Krugman, P. (1991) *Geography and Trade*. MIT Press, Cambridge, MA.
Langhoff, T. (1993) The internationalisation of the firm in an intercultural perspective. MAPP Working Paper No 15, Aarhus School of Business, Aarhus, Denmark.
Marshall A. (1966) *Principles of Economics*. Macmillan, London, p. 731.
Porter, M.E. (1990) *The Competitive Advantage of Nations*. Macmillan, London.

Porter, M.E. (1991) Toward a dynamic theory of strategy. *Strategic Management Journal*, 12, 95–117.

Ritson, C. (1991) The CAP and the consumer, in *The Common Agricultural Policy and the Food Economy* (eds C. Ritson and D. Harvey), CAB International, Wallingford, Oxon, pp. 119–37.

Swinbank, A. and Harris, S. (1991) The CAP and the food industry, in *The Common Agricultural Policy and the Food* (eds C. Ritson and D. Harvey), CAB International, Wallingford, Oxon, pp. 205–19.

Thomas, J., Staatz-John, M. and Pierson-Thomas, R. (1995) Analysis of grocery buying and selling practices among manufacturers and distributors. implications for industry structure and performance. *Agribusiness*, 11(6), November–December, 537–51.

Traill, W.B. and Grunert, K.G. (1997) *Product and Process Innovation in the Food Industry*. Chapman & Hall, London.

Utterback, J. (1974) Innovation in industry and the diffusion of technology. *Science*, 183, 658–62.

Index

Page numbers appearing in **bold** refer to figures and page numbers appearing in *italic* refer to tables.

Printed in the United States
141757LV00001B/2/A

9 780751 404319